D0340774
WITHDRAWN
UTSA LIBRARIES

Canadian Mathematical Society
Société mathématique du Canada

For other titles published in this series, go to
http://www.springer.com/series/4318

Megan Dewar • Brett Stevens

Ordering Block Designs

Gray Codes, Universal Cycles and Configuration Orderings

Megan Dewar
School of Mathematics & Statistics
Carleton University
1125 Colonel By Drive
Ottawa, Ontario
Canada

Brett Stevens
School of Mathematics & Statistics
Carleton University
1125 Colonel By Drive
Ottawa, Ontario
Canada

ISSN 1613-5237
ISBN 978-1-4614-4324-7 ISBN 978-1-4614-4325-4 (eBook)
DOI 10.1007/978-1-4614-4325-4
Springer New York Heidelberg Dordrecht London

Library of Congress Control Number: 2012943379

Mathematics Subject Classification (2010): 05Bxx, 11B50, 05C51, 05C70, 05C45, 94B25, 62K10, 68R15, 68M15, 90B35

Printed on acid-free paper

Springer is part of Springer Science+Business Media (www.springer.com)

To Madeline Victoria North

M.E.D

To Charles Martin Stevens 1920–2011

C.B.S

Preface

This study originated with Brett asking himself "What would a Gray code of a block design be?" during his postdoctoral fellowship at Simon Fraser University in 2000. At that time he was surveying the broad literature on Gray codes for different combinatorial families. Megan began her Ph.D. thesis work on the topic in 2002. At that time we found that we were certainly not the first people to ask this question and that it had, in fact, been asked in many different ways using different terminologies. David Pike, Megan's external thesis reviewer, suggested that we consider turning her work into a book and this document is the result. This study attempts to gather together all these myriad investigations, including our own, in one cohesive whole.

The primary focus of this book is orderings for the blocks of combinatorial designs that mirror Gray codes (i.e. minimal change orderings) and the generalization of this concept to orderings in which every ℓ consecutive blocks have some consistent property. Since these constraints are applied to a block and some small set of its neighbours in a sequence, we think of these as local orderings. We include some discussion of two other kinds of orderings for the blocks of a design: orderings that have some global rather than local property and orderings that optimize some given objective function. These other orderings are not the primary focus of this study. Our purpose is to unify the terminology related to the general question of ordering the blocks of a design, to gather and survey the known results and to review the various methods used by researchers in this area. We also present new results, highlight the open and interesting questions and survey the various applications of block design orderings. We hope that our bibliography will be a useful resource to anyone interested in these and related areas of study.

We begin with an introduction to the ideas and motivations of this study. We follow this with a review of the background mathematical material required in undertaking this study: graph theory, design theory and combinatorial orderings. We then proceed to make connections between these areas by defining various orderings for block designs. These can be roughly divided into two types: configuration orderings where consecutive blocks have structural constraints and minimal change orderings where the changes, although possible to see structurally, are more easily discussed quantitatively. The following chapters survey the known results and proof methods for these two families of orderings. We finish with the diverse uses to which orderings of designs can be applied, ranging over scheduling, reliability and group testing, error correction and statistical experimental design.

We have two intended audiences: mathematical scientists and engineers who are familiar with some of the material and who want to learn more about the connections, methods and applications; and students interested in doing research in this area and who may use the book as a primary resource to teach themselves the relevant background, methods and results, before exploring various open questions.

We would both like to thank Marcia Bunda, Vaishali Damle, Karl Dilcher, Rajiv Monsurate, David Pike, Meredith Rich and the reviewers for their encouragement and assistance in the creation of this book. We would like to thank the many people who helped us ask the right questions and find the diverse results in the literature: Brian Alspach, Joe Buhler, Karel Casteels, Myra Cohen, Charles Colbourn, Persi Diaconis, Jeff Dinitz, Peter Dukes, Peter Eades, Hal Fredricksen, Dalibor Fronček, Luis Goddyn, Fan Chung Graham, Ronald Graham, Mike Grannell, Terry Griggs, Donovan Hare, Peter Horák, Glenn Hurlbert, Garth Isaak, Brad Jackson, David Leach, Brendan McKay, Rudi Mathon, Eric Mendelsohn, Bojan Mohar, Kevin Phelps, David Pike, Alex Rosa, Frank Ruskey, Carla Savage, Doug Stinson, Vadlamudi China Venkaiah, Walter Wallis, Aaron Williams and Peter Winkler.

Megan Dewar would like to thank her partner, Chris, for his support and encouragement, and for bearing his "tech support" role with such equanimity. Most importantly, she would like to acknowledge his skill as a barista which kept her provisioned with lovely lattes throughout her work on this book. She would like to thank John Proos for thorough reading of the thesis from which this document stems—he work greatly improved the original document. Finally, thanks are extended to Brett for being the best thesis supervisor a student could wish for: kind, encouraging,not only interested in the topic but also interested in many other pursuits. Collaborating on this book has been a wonderful culmination of our research together. Brett Stevens would like to thank, first and foremost, Megan for her patience, commitment, intelligence, sense and energy. He would also like to thank his family: Joelle and Greta. Lastly, and in many ways most importantly, he would like to thank Madeline for motivating us to finally finish this book, for usually permitting us to work and for being lovely when not granting that permission.

Ottawa, ON, Canada

Megan Dewar
Brett Stevens

Contents

Chapter 1
Introduction

The mathematicalsciences particularly exhibit order, symmetry and limitation, and these are the greatest forms of the beautiful.
- Aristotle, Metaphysica

In this monograph, we consider the problem of ordering the blocks of designs to meet specified criteria. The problems of ordering binary vectors, k-subsets of n-sets, permutations and other combinatorial objects have been thoroughly explored in the context of combinatorial Gray codes. Further work has been done within the restricted and newer framework of universal cycles (Ucycles). These existing definitions motivate criteria for ordering blocks. Block orderings can be classified into two broad groups: local and global. Our focus will be local orderings, and we will look at various definitions of these orderings (including Gray codes, Ucycles and configuration orderings). We attempt to survey the results in the field, including the translation of results first presented in a different context.

The most common and well-studied ordering concept in combinatorics is the Gray code. Gray codes are well studied from a theoretical standpoint and, in addition, have many practical applications. Gray codes are particularly efficient in testing situations where there is a cost associated to making adjustments to inputs between tests. These costs are frequently a limited resource cost (often time) but sometimes an error tolerance. The name "Gray code" is due to Frank Gray who, in 1953, used a binary reflected code in a pulse code communication system. A binary Gray code of order n lists all 2^n strings of n-bits such that only one bit changes from one string to the next, in a simple and regular way. A binary Gray code is cyclic if the change rule also holds from the last n-bit string to the first n-bit string. In general terms, a combinatorial Gray code for a collection of combinatorial objects is a listing such that successive elements differ in a small, pre-specified way. As Gray codes have both practical applications and appealing properties, their existence, enumeration and generation have been well studied (see Savage's survey paper [41] for an extensive list of papers). While a Gray code for k-subsets of an n-set (with various minimal change properties) is a Gray code for the family of

M. Dewar and B. Stevens, *Ordering Block Designs: Gray Codes, Universal Cycles and Configuration Orderings*, CMS Books in Mathematics,
DOI 10.1007/978-1-4614-4325-4_1, © Springer Science+Business Media New York 2012

BIBD$(n,k,\binom{n-2}{k-2})$s, and the existence of such codes is known [5,7,9,19,39,40,46], Gray codes for designs had not been studied until the work on single-change covering designs of Wallis et al. [44] in the early 1990s. A formal definition of Gray codes for designs was not given until [16].

There are many ways to define minimal change. The minimality may refer to where the change occurs or may refer to how the change occurs. While these two concepts are not mutually exclusive, the standard binary reflected Gray code clearly falls into the first category; the change at each step is specified by the bit position that is to be changed. The ordered list of bit positions that change (known as a transition sequence) completely specifies the code. An example of the second type of change is a de Bruijn sequence. A de Bruijn sequence of order n is a listing of all n-bit strings such that successive strings differ by a shift one position left (dropping the first element) followed by the introduction of a new last element. In other words, this form of change asks that consecutive elements of the de Bruijn sequence be successive states of a queue obtained by a pre-specified series of simultaneous pushes and pops. Here the change at each step is given by the value that is to be pushed onto the queue. This form of change means that successive elements of the sequence may differ radically in each bit position (i.e. successive elements may have large Hamming distance); however, the numbers of zeros and ones appearing may not differ at all. This definition allows us to compress the list of all 2^n n-bit strings into a single cyclic list of 2^n bits.

Formally, a k-ary de Bruijn sequence of order n is a circular sequence of length k^n such that every n-tuple from the alphabet $\mathbb{Z}_k$ (the group structure is not always important) appears as a contiguous subsequence. First investigated by Flye-Sainte Marie in 1894 and later rediscovered by N. de Bruijn, de Bruijn sequences have long been known (Sanskrit memory wheels) and extensively used (fault testing, pseudo-random number generation, magicians' tricks). De Bruijn sequences of order n on alphabets of size k are known to exist for all $n \in \mathbb{N}$ and all $k \in \mathbb{N}$. A natural abstraction of this concept was introduced by Chung, Diaconis and Graham in their 1992 paper "Universal cycles for combinatorial structures" [10]. In this paper they propose representing combinatorial objects (other than the n-tuples on $\mathbb{Z}_k$) in a cyclic sequence such that each object appears exactly once. Formally, suppose $\mathscr{F}_n$ is a family of combinatorial objects of "rank" n and let $m = |\mathscr{F}_n|$. Rank refers to the size of the representative of each object. This term is loosely defined as each type of combinatorial object may be represented differently. For example, k-subsets of an n-set may represent themselves, while partitions of an n-set may be represented by vectors of length n where the value in the ith position indicates the part to which element i belongs. Assume each $F \in \mathscr{F}_n$ is specified by some sequence $x_0, x_1, \ldots, x_{n-1}$, where, for $0 \le i \le n-1$, $x_i \in A$ for some fixed alphabet A. The sequence $U = a_0, a_1, \ldots, a_{m-1}$ is a universal cycle (Ucycle) for $\mathscr{F}_n$ if $a_{i+1}, \ldots, a_{i+n}$, $0 \le i < m$, runs through each element of $\mathscr{F}_n$ exactly once, where index addition is performed modulo m. The paper sparked a flurry of work which produced results on the existence of Ucycles for k-subsets of an n-set, permutations of an n-set and partitions of an n-set. The existence of Ucycles for k-permutations of an n-set, $2 < k < n$, is completely determined [28] as is the existence of Ucycles for partitions of an n-set [10]. Johnson [30] has proved that an alphabet of $n+1$ symbols is sufficient for order-isomorphic Ucycles for n-permutations of an n-set. However, the existence of Ucycles for k-subsets of an n-set is far from complete. It is known that Ucycles for k-subsets of an n-set, where $k = 3,4,6$, exist when n is sufficiently large and coprime to k, and provided $\binom{n-1}{k-1} \equiv 0 \pmod{k}$ [26–28]. Notice that a Ucycle for subsets exhibits both minimal change properties; consecutive elements differ by a shift and also differ by one element (since order of elements in a subset is irrelevant). We can think of Ucycles for subsets as minimal change listings in which the number that has been "in the subsets" longest is changed. This is similar in flavour to the definition of a Beckett Gray code [15].

Given these results, as design theorists, we naturally wonder whether minimal change orderings exist for block designs. Until 2007, this question had not been posed in these terms; however, we will see that work had already been done on answering this question in some specific cases. The natural questions to ask are: how do we define sequences of blocks that are Gray codes or Ucycles? Are there designs which admit these sequences? Do these sequences exist for all designs? If so, how many are there and how do we construct them? In order to answer these questions, we first approach definitions. A Gray code type order is quite natural to apply to a family of sets like the blocks of a combinatorial design; however, this can be done in one of two ways. Given a block design we seek to construct an ordering of the blocks so that the size of intersection between consecutive blocks is as large as possible; in other words, the minimum number of points changes between consecutive blocks. When the design has index one then the overlap is size at most one and the change is unavoidably large, but, as the index increases, "better" minimal change is possible. Alternatively, we can insist that between two consecutive blocks of size k, only one element changes. Unless the index is sufficiently high this is impossible for many designs so a natural accommodation is to look for the smallest covering design existing with this property. This was investigated in the 1990s under the name "single-change covering designs". The former definition has been considered in terms of Hamiltonicity of block-intersection graphs and in some direct constructions. Of course, some designs admit orderings which are examples of both. For example, Dewar proved that there exists a twofold triple system (TTS(v)), for each $v \equiv 3 \pmod{12}$ that admits a cyclic Gray code in which consecutive objects share two points [16]. This is called a 2-intersecting cyclic Gray code and is also an example of single-change.

In contrast, Ucycles (of any family of objects) must address how an object is to be represented as a string. Unlike de Bruijn sequences, there are many ways to represent the combinatorial objects in question and how objects are represented affects whether Ucycles can be found. We define Ucycles for block designs and look at two realizations of this definition. In one realization, blocks of the design in question are represented based on their cardinality or block size, while in the other realization, representation is based on the strength of the design. First, we look at twofold triple systems of order v, which exist for all $v \equiv 0,1 \pmod{3}$. Dewar proved there exists a TTS(v) for each $v \equiv 1,4,7 \pmod{12}$, with sporadic exceptions,

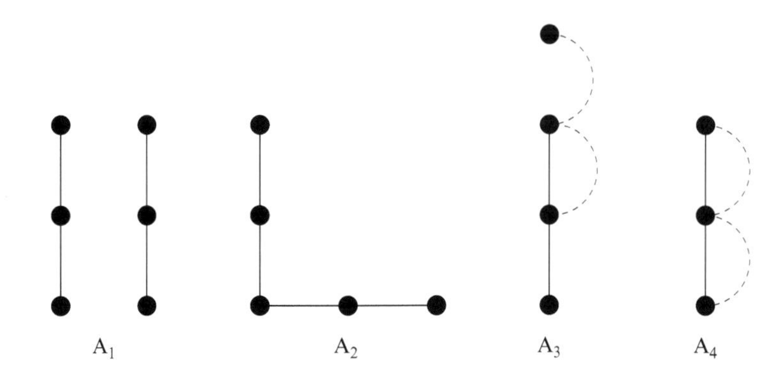

Fig. 1.1 Configurations on two lines (labels from [13] and [20])

that admits a Ucycle of rank three. These are also single-change orderings. Here the rank is the block size and each block represents itself in the Ucycle. When $v \equiv 0,3,6,9 \pmod{12}$, she showed that Ucycles of rank three for TTS(v)s cannot exist [16]. She proved by construction that Ucycles of rank two exist for all cyclic BIBDs having a sufficient number of non-regular base blocks [16]. In these Ucycles, the rank is the strength (2), and blocks are represented by a pair of points on them, much like a vector space can be represented by a basis. Techniques often seen in Gray code construction are applied to the construction of Ucycles for block designs, indicating that Ucycles for designs are a natural extension of this concept.

The concept of ordering the blocks of a design is not limited to Gray codes and Ucycles. In 2003, Cohen and Colbourn introduced the idea of a configuration ordering for the blocks of a design [11]. A (p,ℓ)-configuration is a set system with p elements and ℓ blocks in which every element is contained in at least one block. Let C be a configuration having ℓ blocks. A configuration ordering, or C-ordering, for a design is a listing of the blocks of the design such that every ℓ consecutive blocks form a configuration isomorphic to C. If this property also holds when we treat the list as a cycle, then the ordering is called C-cyclic. Cohen and Colbourn asked, for various configurations C, when does there exist a Steiner triple system of order v (STS(v)) for which the triples can be C-ordered? Of course, this question can be asked of all triple systems and block designs in general.

Configurations are easiest to express visually. Figures 1.1 and 1.2 represent some of the configurations on blocks of size three that we discuss in this book. We employ the labelling of configurations given by Colbourn and Rosa in [13] and Griggs and Grannell in [20]. Note that specifying the number of blocks and the number of points may not uniquely determine a configuration. For example, B_3 and B_4 (Fig. 1.2) are both $(7,3)$-configurations. In such cases, we will fully describe the configuration in question. As defined above, configurations need not have blocks of size three; however, there is no standard notation that describes these more general configurations. Given a configuration C with blocks of size three, we will use the notation C' to refer to the generalized version of this configuration, describing the generalization in detail where necessary.

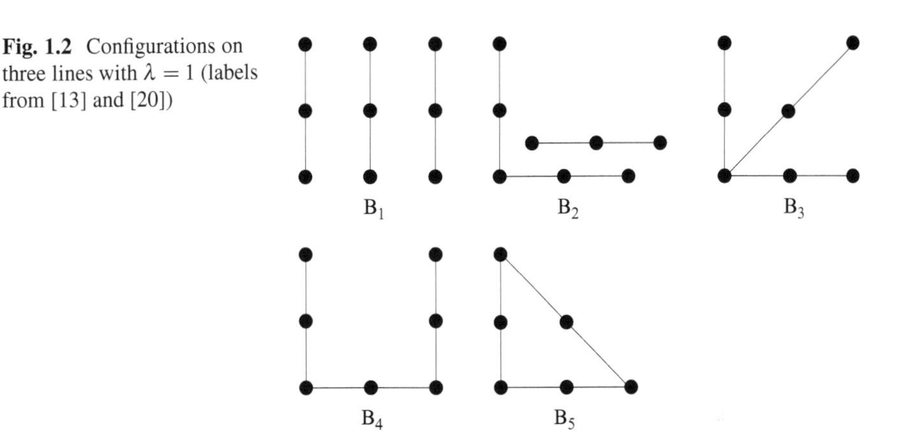

Fig. 1.2 Configurations on three lines with $\lambda = 1$ (labels from [13] and [20])

While few people have looked directly at existence of configuration orderings, many known results can be expressed in this language. Simmons and Davis proved that all BIBD$(v,2,1)$s admit $(2d,d)$-orderings if and only if $d \leq \lfloor (v-3)/2 \rfloor$ [42]. Note that for any $m,k \in \mathbb{N}$, the $(m \cdot k,k)$-configuration consists of k mutually disjoint blocks of size m. Momihara and Jimbo, motivated by a group testing application discussed in Sect. 6.4, proved the existence of $(6,2)$-cyclic orderings (A_1-cyclic orderings (see Fig. 1.1)) for maximum packings by triples and $(8,2)$-cyclic orderings for some Steiner quadruple systems [33, 34]. Horák and Rosa proved that every BIBD$(v,k,1)$ admits an A_2'-cyclic ordering (see Fig. 1.1) by proving that the 1 block-intersection graph of every BIBD$(v,k,1)$ is Hamiltonian [25]. The i-block-intersection graph of a design $S = (V, \mathscr{B})$ is the graph formed by taking blocks as vertices and connecting two vertices if the blocks they represent intersect in exactly i points. Several groups of authors have enhanced Horák and Rosa's result. Horák, Pike and Raines showed that all triple systems of order greater than or equal to twelve admit A_2-cyclic orderings [24], while Jesso proved that the 1 block-intersection graph of every BIBD$(v,4,\lambda)$, for $v \geq 136$, and the 1 block-intersection graph of every BIBD$(v,5,\lambda)$, for $v \geq 305$, is Hamiltonian [29]. The work of Alspach, Heinrich and Mohar implies that any PBD$(v,K,1)$ (where K represents a collection of integers, each less than v) with $\max(K) \leq 2 \cdot \min(K)$ admits a cyclic configuration ordering in which pairs of consecutive blocks intersect in exactly one point [1]. Hare proved that the 1 block-intersection graph of every PBD$(v,K,1)$ with $\min(K) \geq 3$ is edge-pancyclic, a result which implies that such 1 block-intersection graphs are Hamiltonian, and hence every PBD$(v,K,1)$ with $\min(K) \geq 3$ admits an ordering in which pairs of consecutive blocks (of any size) intersect in exactly one point [22]. Recently, Case and Pike have shown that every PBD(v,K,λ), with $\lambda \geq 2$ and $\max(K) \leq \lambda \min(K)$, admits a cyclic ordering in which every pair of consecutive blocks intersects (perhaps in multiple points) [8]; this generalizes the work of Mamut, Pike and Raines [33]. A similar result involving Teirlinck designs was proved by Pike, Vandell and Walsh [36]. On the other hand, M. Colbourn and Johnstone have dealt with the non-existence of configuration orderings for triple systems. In [14], they present a simple TTS(v)

that does not admit an A_3-ordering (see Fig. 1.1). They imply that there exists a family of simple twofold triple systems having this property and give a potential method for constructing them.

Cohen and Colbourn's work on configuration orderings stems from an erasure-correction application. Many computer architectures connect a large number of disks together in an attempt to improve I/O performance. A disk fails when some bits of the disk are deemed unreadable; an unreadable bit is called an erasure. Erasure-correcting codes are designed to recover erased bits of data when the location of the erased bits is known. Given an erasure-correcting code, Cohen and Colbourn use orderings of the blocks of STSs to represent orderings of disks that are either good or bad for system performance. Cohen and Colbourn have shown that for all admissible $v \geq 9\ell - 6$, $\ell \in \mathbb{N}$, there exists an STS(v) admitting a $(3\ell, \ell)$-ordering [11]. Further, for $\ell \in \mathbb{N}$ and admissible $v \geq 81(\ell - 1) + 1$, *every* STS(v) admits a $(3\ell, \ell)$-ordering [11]. They have also shown that for all admissible $v \geq 15$, there exists an STS(v) that is not B_5-orderable (see Fig. 1.2) [11].

Configuration orderings are interesting in their own right, and they also provide a useful framework in which to discuss other ordering concepts. In particular, we can frame questions about the existence of single-change orderings, Ucycles and Gray codes in terms of the existence of a particular configuration ordering. We will see that at times this is not an enlightening shift in perspective; however, in many cases it has proved useful. As configuration ordering is an emerging area of study, few configurations have been investigated, and the investigation has focused on Steiner triple systems and complete graphs (designs of block size two). Dewar expanded the known results regarding existence of A_1- and B_1-cyclic orderings to all triple systems [16], as Cohen and Colbourn's results deal exclusively with STSs. Single-change covering designs are minimum covers which have an A_3-ordering. As mentioned earlier, Dewar proves that there exists either a Ucycle of rank three or a 2-intersecting cyclic Gray code for the blocks of at least one TTS(v) for each order $v \equiv 1,3,4,7 \pmod{12}$, with sporadic exceptions [16]. This result implies the existence of an A_3-cyclic ordering for these designs. For triple systems with $\lambda = 2$, this is a good general result, as the authors of [14] suggest that there exists an infinite family of TS(v,λ)s (but only show the existence of one specific design) that do not admit such orderings. The question of existence of A_3-orderings remains open for TS(v,λ)s with $\lambda \geq 3$. An A_4-ordering (see Fig. 1.1) exists only for the trivial BIBD$(3,3,\lambda)$s where $\lambda \geq 2$; thus, the next configuration that is interesting is B_2—known as the hut configuration (see Fig. 1.2). Dewar has proved that every TS(v,λ), $v \geq 137$, admits a hut-cyclic ordering [16].

Configuration orderings of triple systems can be generalized in two different ways. Configurations can be extended to block size $k > 3$ and ordering results for BIBD$(v,k,1)$s sought. For example, let A_1' denote a $(2k,2)$-configuration, let B_1' denote a $(3k,k)$-configuration and let B_2' denote the configuration of three blocks with two blocks intersecting in a single point and the third disjoint. Dewar has proved that every BIBD$(v,k,1)$, $v \geq 2k^2 + 1$, admits an A_1'-cyclic ordering, that every BIBD$(v,k,1)$, $v \geq 12k^2+1$, admits a B_1'-cyclic ordering, and that every BIBD$(v,k,1)$, $v \geq 18k^2-6k+1$, admits a B_2'-cyclic ordering [16].

In [16], the concept of single configuration ordering was generalized. Let $\mathscr{C}$ be a set of configurations, each having ℓ blocks. A $\mathscr{C}$-ordering for a design is a listing of the blocks of the design such that every ℓ consecutive blocks form a configuration isomorphic to one of the configurations in $\mathscr{C}$. When the list is cyclic, the ordering is called a $\mathscr{C}$-cyclic ordering. To distinguish between the two definitions of configuration ordering, we will call $\mathscr{C}$-orderings generalized orderings, while a configuration ordering as defined by Cohen and Colbourn will be called a standard ordering. The definition of generalized configuration ordering allows us to discuss all block ordering results in a common language. For example, a Ucycle of rank two for a BIBD(v,k,λ) is equivalent to a $\mathscr{C}$-cyclic ordering, where $\mathscr{C}$ is a set of eight different configurations on three blocks. Again, most work on orderings of designs has not been stated in these terms but is readily expressible; for example, Hamilton cycles in the $\{1,2\}$ block-intersection graph of a triple system are equivalent to $\{A_2,A_3\}$-cyclic orderings. Similar translation can be done for results on the Hamiltonicity of block-intersection graphs for designs with $k>3$. Ucycles for the TS(v)s previously discussed in this chapter are $\{B_6,B_7\}$-cyclic orderings. Similar to the standard configuration ordering case, most of the work on generalized configuration ordering has focused on smaller configurations with less than four blocks. In particular, consideration has been given to the existence of $\{B_4,B_5\}$-orderings. B_4 and B_5 are called the 3-path and the triangle, respectively. There are two motivations for determining the existence of $\{B_4,B_5\}$-cyclic orderings of the blocks of designs. The first is that when $\lambda=1$, a $\{B_4,B_5\}$-cyclic ordering is exactly a 1-intersecting cyclic Gray code. That is, the $\{B_4,B_5\}$-cyclic ordering is a minimal change ordering for the blocks of a design because any two blocks from such a design intersect in at most one point. While results regarding the Hamiltonicity of the 1 block-intersection graph have already proved the existence of such designs, a $\{B_4,B_5\}$-ordering has the additional property that no point stays on consecutive blocks of the list for "too long" (a Ucycle-like property). Every cyclic STS(v), $v \neq 3$, admits a $\{B_4,B_5\}$-cyclic ordering. Since for each admissible order there exists a cyclic STS(v), this result implies that for each admissible order $v \neq 3$, there exists an STS(v) that admits a $\{B_4,B_5\}$-cyclic ordering. This ordering is best possible in the sense that Cohen and Colbourn have shown that there exists a Steiner triple system of every admissible order that does not admit a B_5-ordering [11]. The second motivation for considering the existence of $\{B_4,B_5\}$-cyclic orderings relates to a design theoretic analogue of a graph theoretic concept. It is easy to show that if a graph is Eulerian, its line graph is Hamiltonian. Furthermore, it is easy to find an example proving that the converse of this statement is false. This leads to the following well-known theorem: a graph G is Eulerian if and only if its line graph is Hamiltonian and no pair of consecutive edges in the Hamilton cycle represents a 3-claw in G [45]. The analogue of this theorem in design theory requires the block-intersection graph. The block-intersection graph of a design $S=(V,\mathscr{B})$ is a graph with vertices representing the blocks of $\mathscr{B}$ and two vertices, B_1 and B_2, connected if $B_1 \cap B_2 \neq \emptyset$. We associate the block-intersection graph of a design with the line graph of a graph. By analogy to a graph, a design is Eulerian if and only if the block-intersection graph is Hamiltonian and there exists a Hamilton cycle in the block-intersection graph which does not pass through any 3-claws in the design. Note that this definition of Eulerian is not the same as one used in the context of hypergraphs [3]. The existence of Hamilton cycles in block-intersection graphs has been settled for several large families of designs [1, 24]—are these designs Eulerian? That is, does there exist a Hamilton cycle in the block-intersection graph having the required property? The restriction on the Hamilton cycle can easily be expressed using configuration ordering terminology by explicitly not allowing the claw configuration on three blocks to appear.

The concepts of minimal change and configuration ordering can be extended to graph decomposition. Graph decompositions are partitions of the complete graph into edge disjoint copies of some fixed subgraph (BIBDs are partitions of the complete graph into smaller complete graphs (K_k)). The decompositions we consider herein are primarily matchings, cycles and disjoint collections of cycles. We see both minimal change orderings, as in single-change neighbour designs (which are decompositions into cycles) and configuration orderings, as in sequentially perfect 1-factorizations (where the requirement on consecutive matchings is structural: their union must form a Hamilton cycle). Frequently graph decomposition problems are motivated by specific applications which determine the fixed subgraph for the partition. For example, tournament scheduling problems call for decompositions into partial matchings.

Before immersing ourselves in the subject, we must emphasize what this book is *not* about. The statement "ordering the blocks" could be interpreted to mean ordering the points within each block. This is not what the book is about; however, we acknowledge these are themselves interesting objects and refer the interested reader to material where she can start learning about them. Ordered designs of this type include:

- Bridge tournaments [17, 18]
- Complete Latin squares and balanced arrays [43]
- Court-balanced tournament designs [18, 31, 38]
- Cycle systems [6]
- Directed designs [4]
- Linearly ordered block designs [43]
- Mendelsohn designs [35]
- Mixed doubles tournaments [18]
- Neighbour designs (cycle systems where any point may appear multiple times in the cycle, but not beside itself) [43]
- Ordered orthogonal arrays [34]
- Polyhedral designs [21, 23, 32]
- Regular r-tournaments [18]
- Whist tournaments [2, 18]
- Youden squares [37].

Some designs combine the notion of ordering points within blocks *and* ordering the blocks themselves, namely, home and away balanced tournaments, directional seed orchard designs and court-balanced tournament designs which are also interval balanced. We will discuss these in this monograph, but our attention will be focused on the ordering of the blocks. We refer the reader who is interested in the ordering of points within blocks in home and away balanced tournaments to [18, 38].

Finally, note that the orderings we have discussed in this introduction involve sets of s consecutive blocks in the ordering having some local property; cardinality of the changes or isomorphism to a configuration. We have encountered two other notions of ordering the blocks of a design in the literature. We will discuss these orderings, although they are not the primary focus of the book. The first notion is an ordering of the blocks where the set of all s consecutive blocks induces a set system with particular properties. For example, the collection of unions of each s consecutive blocks are distinct or the set system of the unions is a design of some kind in and of itself. Another way of defining orderings of the blocks of a design is to seek to maximize some objective function which is defined for various block orders. Both of these concepts will be defined formally in Sect. 3.3, and we will discuss them primarily in terms of their applications in Chap. 6.

The remainder of this monograph is organized as follows. Chapter 2 presents some basic terminology and results in graph and design theory, in the process introducing symbols that will be used throughout the book. Chapter 2 also provides background on Gray codes and Ucycles. In Chap. 3, we formally define the orderings for block designs considered in this monograph. Chapters 4 and 5 are the main body of the survey. Chapter 4 deals with configuration orderings, while Chap. 5 looks at orderings which are best defined as minimal change. In Chap. 6 we discuss various applications of ordering the blocks of designs—many of these applications motivated the initial investigations of these concepts. Throughout, chapters end with a conjectures section (where relevant) and an exercises and open problems section. Exercises are intended to be pedagogical, allowing the reader to develop their familiarity with the definitions and methods. Problems are open research problems of significant interest. We note that each chapter contains a list of references used within it. While this makes the chapters more self-contained, the same reference may have different numbers in different chapters and the same citation number in two different chapters is likely to refer to two different sources.

References

1. Alspach, B., Heinrich, K., Mohar, B.: A note on Hamilton cycles in block-intersection graphs. In: Finite Geometries and Combinatorial Designs (Lincoln, NE, 1987), Contemporary Math, vol. 111, pp. 1–4. American Mathematical Society, Providence, RI (1990)
2. Anderson, I., Finizio, N.J.: Whist Tournaments, chap. VI.64, pp. 663–668. In: Colbourn and Dinitz [12] (2007)
3. Batzoglou, S., Istrail, S.: Physical mapping with repeated probes: the hypergraph superstring problem. J. Discrete Algorithms (Oxf.) **1**(1), 51–76 (2000)

4. Bennett, F.E., Mahmoodi, A.: Directed Designs, chap. VI.20, pp. 441–444. In: Colbourn and Dinitz [12] (2007)
5. Bitner, J.R., Ehrlich, G., Reingold, E.M.: Efficient generation of the binary reflected Gray code and its applications. Comm. ACM **19**(9), 517–521 (1976)
6. Bryant, D., Rodger, C.: Cycle Decompositions, chap. VI.12, pp. 373–382. In: Colbourn and Dinitz [12] (2007)
7. Buck, M., Wiedemann, D.: Gray codes with restricted density. Discrete Math. **48**, 163–171 (1984)
8. Case, G.A., Pike, D.A.: Pancyclic PBD block-intersection graphs. Discrete Math. **308**, 896–900 (2008)
9. Chase, P.J.: Combination generation and Graylex ordering. Congr. Numer. **69**, 215–242 (1989)
10. Chung, F., Diaconis, P., Graham, R.: Universal cycles for combinatorial structures. Discrete Math. **110**, 43–59 (1992)
11. Cohen, M.B., Colbourn, C.J.: Optimal and pessimal orderings of Steiner triple systems in disk arrays. Theoret. Comput. Sci. **297**, 103–117 (2003)
12. Colbourn, C.J., Dinitz, J.H. (eds.): Handbook of Combinatorial Designs, second edn. Chapman & Hall/CRC, Boca Raton, FL (2007)
13. Colbourn, C.J., Rosa, A.: Triple Systems. Oxford Mathematical Monographs. The Clarendon Press Oxford University Press, New York (1999)
14. Colbourn, M.J., Johnstone, J.K.: Twofold triple systems with a minimal change property. Ars Combin. **18**, 151–160 (1984)
15. Cooke, M., Dewar, M., North, C., Stevens, B.: Beckett Gray codes. Submitted to J. Combin. Math. Combin. Comput. (2009)
16. Dewar, M.: Gray codes, universal cycles and configuration orderings for block designs. Ph.D. thesis, Carleton University, Ottawa (2007)
17. Dinitz, J.H.: Howel Designs, chap. VI.29, pp. 499–504. In: Colbourn and Dinitz [12] (2007)
18. Dinitz, J.H., Frončeк, D., Lamken, E.R., Wallis, W.D.: Scheduling a Tournament, chap. VI.51, pp. 591–606. In: Colbourn and Dinitz [12] (2007)
19. Eades, P., Hickey, M., Read, R.C.: Some Hamilton paths and a minimal change algorithm. J. Assoc. Comput. Mach. **31**(1), 19–29 (1984)
20. Grannell, M.J., Griggs, T.S.: Configurations in Steiner triple systems. In: Combinatorial Designs and their Applications (Milton Keynes, 1997), vol. 403, pp. 103–126. Chapman & Hall/CRC, Boca Raton, FL (1999)
21. Hanani, H.: Decomposition of hypergraphs into octahedra. In: Second International Conference on Combinatorial Mathematics (New York, 1978), Annals of the New York Academy of Sciences, vol. 319, pp. 260–264. New York Academy of Sciences, New York (1979)
22. Hare, D.R.: Cycles in the block-intersection graph of pairwise balanced designs. Discrete Math. **137**, 211–221 (1995)
23. Hartman, A., Phelps, K.T.: The spectrum of tetrahedral quadruple systems. Utilitas Math. **37**, 181–188 (1990)
24. Horák, P., Pike, D.A., Raines, M.E.: Hamilton cycles in block-intersection graphs of triple systems. J. Combin. Des. **7**(4), 243–246 (1999)
25. Horák, P., Rosa, A.: Decomposing Steiner triple systems into small configurations. Ars Combin. **26**, 91–105 (1988)
26. Hurlbert, G.H.: Universal cycles: on beyond de Bruijn. Ph.D. thesis, Rutgers University, New Brunswick, NJ (1990)
27. Hurlbert, G.H.: On universal cycles for k-subsets of an n-set. SIAM J. Disc. Math. **7**(4), 598–604 (1994)
28. Jackson, B.W.: Universal cycles of k-subsets and k-permutations. Discrete Math. **117**, 141–150 (1993)
29. Jesso, A.: The Hamiltonicity of block-intersection graphs. Master's thesis, Memorial University of Newfoundland, St. John's, NF (2010)
30. Johnson, J.R.: Universal cycles for permutations. Discrete Math. **309**, 5264–5270 (2009)

31. Lamken, E.R.: Balanced Tournament Designs, chap. VI.3, pp. 333–336. In: Colbourn and Dinitz [12] (2007)
32. Linek, V., Stevens, B.: Octahedral designs. J. Combin. Des. **18**(5), 319–327 (2010)
33. Mamut, A., Pike, D.A., Raines, M.E.: Pancyclic BIBD block-intersection graphs. Discrete Math. **284**, 205–208 (2004)
34. Martin, W.J.: (t,m,s)-Nets, chap. VI.59, pp. 639–643. In: Colbourn and Dinitz [12] (2007)
35. Mendelsohn, E.: Mendelsohn Designs, chap. VI.35, pp. 528–534. In: Colbourn and Dinitz [12] (2007)
36. Pike, D.A., Vandell, R.C., Walsh, M.: Hamiltonicity and restricted block-intersection graphs of t-designs. Discrete Math. **309**, 6312–6315 (2009)
37. Preece, D.A., Colbourn, C.J.: Youden Squares and Generalized Youden Designs, chap. VI.65, pp. 668–674. In: Colbourn and Dinitz [12] (2007)
38. Rodney, P.: Balance in tournament designs. Ph.D. thesis, University of Toronto, Toronto, ON (1993)
39. Ruskey, F.: Adjacent interchange generation of combinations. J. Algorithms **9**(2), 162–180 (1988)
40. Ruskey, F.: Simple combinatorial Gray codes constructed by reversing sublists. In: Proceedings of the Fourth International Symposium (ISAAC '93) held in Hong Kong, Dec 15–17, 1993. Edited by K.W. Ng, P. Raghavan, N.V. Balasubramanian and F.Y.L. Chin. Lecture Notes in Computer Science, vol. 762, Springer-Verlag, Berlin, pp. xiv+542 (1993). ISBN: 3-540-57568-5
41. Savage, C.D.: A survey of combinatorial Gray codes. SIAM Rev. **39**(4), 605–629 (1997)
42. Simmons, G.J., Davis, J.A.: Pair designs. Comm. Statist. **4**, 255–272 (1975)
43. Street, A.P., Street, D.J.: Combinatorics of Experimental Design. Oxford Science Publications. The Clarendon Press Oxford University Press, New York (1987)
44. Wallis, W., Yucas, J.L., Zhang, G.H.: Single-change covering designs. Des. Codes Cryptogr. **3**, 9–19 (1992)
45. West, D.B.: Introduction to Graph Theory. Prentice Hall, Upper Saddle River, NJ (2001)
46. Wilf, H.S.: Combinatorial algorithms: an update. CBMS-NSF Regional Conference Series in Applied Mathematics, 55. Society for Industrial and Applied Mathematics (SIAM), Philadelphia, PA (1989)

Chapter 2
Background

This chapter provides the basic background necessary for understanding subsequent chapters. We begin with an introduction to the graph theory, designs and design theory concepts used throughout this monograph. It is not required reading for someone with knowledge of graph and/or design theory; however, we formalize the symbols and terminology that will be used throughout. Design theoretic definitions come from [51], unless otherwise indicated. The second half of the chapter deals with more advanced concepts, including configurations, Gray codes and universal cycles (Ucycles). For a more detailed overview of these concepts and a history of major results in each area, see [10, 15, 23, 44].

2.1 Graph Theory

In this section, we define only the graph theory terminology required in later parts of this monograph. This is by no means a complete introduction to the subject. For a broad introduction to graph theory, see [56]. Define a **graph** $G = (V, E)$, where V is a set of vertices and E is a set of edges having endpoints in V. Often, given a graph G, the notation $V(G)$ (respectively, $E(G)$) is used to indicate the vertex set of G (respectively, edge set). A **simple** graph is a graph having no loops and no repeated edges. The **degree** of a vertex $v \in V$ is the number of edges incident with v. By convention, $\delta(G)$ denotes the minimum vertex degree of G, and $\Delta(G)$ denotes the maximum vertex degree of G. A **walk** in G is an alternating list of vertices and edges, $v_0, e_1, v_1, \ldots, e_k, v_k$, such that for $1 \leq i \leq k$, the endpoints of e_i are v_{i-1} and v_i. A **trail** is a walk with no repeated edge, and a **path** is a walk with no repeated vertex. A walk, trail or path is **closed** if the two endpoints, v_0 and v_k, are the same. A closed trail is sometimes called a **tour**, and a closed path is sometimes called a **cycle**. A graph G is **edge-pancyclic** if for every edge $e \in E(G)$ and every integer n, $3 \leq n \leq |V(G)|$, there is a cycle of length n in G containing e.

A **Hamilton path** is a path that contains every vertex of a graph exactly once. Similarly, a **Hamilton cycle** is a cycle that contains every vertex of a graph

M. Dewar and B. Stevens, *Ordering Block Designs: Gray Codes, Universal Cycles and Configuration Orderings*, CMS Books in Mathematics,
DOI 10.1007/978-1-4614-4325-4_2, © Springer Science+Business Media New York 2012

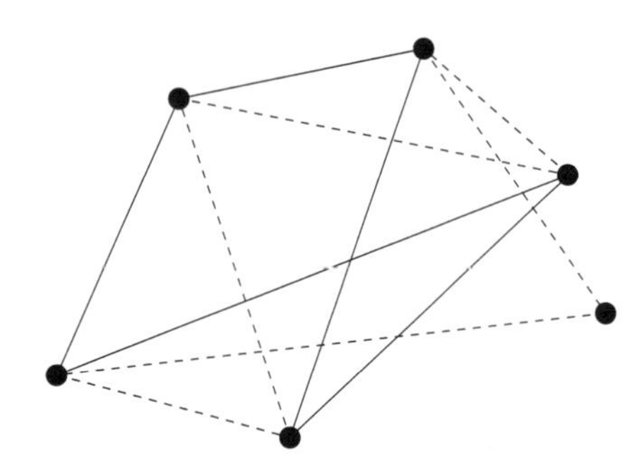

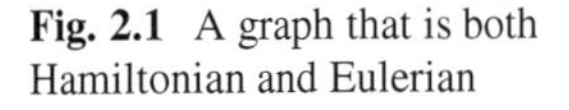

Fig. 2.1 A graph that is both Hamiltonian and Eulerian

exactly once. A graph having a Hamilton cycle is called **Hamiltonian**. A graph is said to be **Hamilton-connected** if the graph has a Hamilton path between every pair of vertices. That is, there exists a path between each pair of vertices which passes through all other vertices. An **Euler tour** is a tour that uses each edge of a graph exactly once; a graph exhibiting such a tour is called **Eulerian**. The graph in Fig. 2.1 is both Hamiltonian and Eulerian. A Hamilton cycle is indicated by dotted edges.

In this monograph, we will make use of the following theorems regarding Hamiltonicity.

Theorem 2.1 (See [56]). *Let G be a simple graph on n vertices with vertex degrees* $d_1 \leq \cdots \leq d_n$*, where* $n \geq 3$*. If* $i < n/2$ *implies that* $d_i > i$ *or* $d_{n-i} \geq n-i$*, then G is Hamiltonian.*

Theorem 2.2 (See [56]). *If* $G = (V,E)$ *has* $|V| \geq 3$ *and* $\delta(G) \geq |V|/2$*, then G is Hamiltonian.*

Given a graph $G = (V,E)$ and a subset of vertices $W \subseteq V$, the subgraph of G induced by the set W is denoted $G[W]$. A **supergraph** of G, denoted $G' = (V',E')$, is a graph with $V \subseteq V'$ and $E \subseteq E'$. A **cutset** of a graph G is a set of vertices whose removal from the graph (along with their associated edges) disconnects the graph. That is, $C \subset V$ is a cutset of G if the graph on $G[V \setminus C]$ is disconnected. A graph is said to be ***s*-connected** if every cutset contains at least s vertices. An **independent set** is a set of vertices that have no edges between them. The **vertex independence number** of a graph G, denoted $\alpha(G)$, is the cardinality of the largest independent set.

Another useful theorem regarding Hamiltonicity is the following:

Theorem 2.3 ([11]). *Let G be a graph with at least three vertices. If, for some s, G is s-connected and contains no independent set of more than s vertices, then G has a Hamilton cycle. If the connectivity is strictly less than the independence number, then the graph is Hamilton-connected.*

The **line graph** of a graph $G = (V,E)$ is the graph with vertex set E and an edge between vertices e_1 and e_2 if the edges in G represented by these vertices share a common endpoint.

The notation K_v denotes the **complete graph** on v vertices, that is, the graph on v vertices having an edge between every pair of vertices. A **bipartite** graph is one in which the vertex set can be partitioned into two parts, such that each edge has one vertex in each part. The partitioned vertex set is commonly identified as $V = X \cup Y$ and the graph is denoted $G = (X,Y,E)$. The following theorem is due to Chvátal.

Theorem 2.4 (See [56]). *Let $G = (X,Y,E)$ be a simple bipartite graph with $|X| = |Y| = v/2 > 1$. Let G' be the supergraph of G such that $G'[Y] = K_{v/2}$. G is Hamiltonian if and only if G' is Hamiltonian.*

A **digraph** is a graph having directed edges. A **2-in 2-out graph** is a digraph in which every vertex has in-degree two and out-degree two. A **strongly connected** digraph is a digraph in which there is a directed path between any two vertices. The following theorem will be appealed to regularly in the construction of de Bruijn sequences and Ucycles.

Theorem 2.5 (See [5]). *Let D be a strongly connected digraph. D has a directed Euler tour if and only if the in-degree of each vertex equals its out-degree.*

A **proper *k*-colouring** of a graph G is a labelling $f : V(G) \to S$, with $|S| = k$, such that adjacent vertices have different labels. The **chromatic number**, $\chi(G)$, is the least k such that G has a proper k-colouring. The greedy algorithm shows that $\chi(G) \leq \Delta(G) + 1$ and those graphs meeting this bound have been precisely characterized.

Theorem 2.6 (Brookes' theorem; see [56]). *If G is a connected graph other than a complete graph or an odd cycle, then*

$$\chi(G) \leq \Delta(G).$$

Finally, an ***r*-factor** of a graph G is a subgraph of G in which every vertex has degree r. An ***r*-factorization** of G is an edge disjoint decomposition of G into r-factors. A 1-factorization of a graph is also called a decomposition into **perfect matchings**—we will use this terminology interchangeably. A **partial matching** is a subgraph of G such that the degree of each vertex is at most one. To refer to a partial matching with precisely c edges, we will use ***c*-matching**.

2.2 Designs

2.2.1 Balanced Incomplete Block Designs

In this section, we define the designs that will be the focus of discussion in the following chapters.

Definition 2.1 (design). A design, or set system, is a pair $(V, \mathscr{B})$ such that V is a set of elements called points and $\mathscr{B}$ is a collection of non-empty subsets of V called blocks.

Table 2.1 A 2-$(7,3,1)$ design

$\{0,1,3\}$	$\{3,4,6\}$	$\{0,2,6\}$
$\{1,2,4\}$	$\{0,4,5\}$	
$\{2,3,5\}$	$\{1,5,6\}$	

Table 2.2 A 3-$(8,4,1)$ design, also denoted SQS(8)

$\{0,1,3,7\}$	$\{3,4,6,7\}$	$\{0,2,6,7\}$
$\{1,2,4,7\}$	$\{0,4,5,7\}$	$\{2,3,5,7\}$
$\{1,5,6,7\}$	$\{2,4,5,6\}$	$\{0,1,2,5\}$
$\{1,3,4,5\}$	$\{0,3,5,6\}$	$\{1,2,3,6\}$
$\{0,1,4,6\}$	$\{0,2,3,4\}$	

Definition 2.2 (t-(v,k,λ) design). Let t,v,k and λ be positive integers such that $2 \leq t \leq k \leq v$. A t-(v,k,λ) design is a set system $(V,\mathscr{B})$, such that the following properties hold: (1) $|V| = v$, (2) each block contains exactly k points, and (3) every t-set of V is contained in exactly λ blocks.

For example, the blocks of a 2-$(7,3,1)$ design are listed in Table 2.1.

When t, known as the **strength** of the design, is equal to two, a t-(v,k,λ) design is called a **balanced incomplete block design** and is denoted BIBD(v,k,λ). Many of the results presented in this monograph focus on **triple systems**, BIBD(v,k,λ)s having $k = 3$. A triple system on v points is denoted TS(v,λ). When $\lambda = 1$, the triple system is called a **Steiner triple system** and is denoted STS(v). When $\lambda = 2$, the triple system is called a **twofold triple system** and is denoted TTS(v). When $t = 3$, $k = 4$ and $\lambda = 1$, a t-(v,k,λ) design is called a **Steiner quadruple system** and denoted is SQS(v). Table 2.2 gives the blocks of an SQS(8).

It is standard practice to let b denote the number of blocks in a given design, and to let r_x represent the number of blocks in which the point x appears. When every point of the design appears in the same number of blocks, simply denote this **replication number** by r.

Theorem 2.7 (See [51]). *A BIBD(v,k,λ), $S = (V,\mathscr{B})$, has exactly*

$$b = \frac{v(v-1)\lambda}{k(k-1)}$$

blocks. Moreover, each element of V is contained in exactly

$$r = \frac{(v-1)\lambda}{k-1}$$

blocks.

The property represented by λ is known as the **index** of the design. It is sometimes useful to express all five parameters of a BIBD, in which case we write BIBD(v,b,r,k,λ). A basic requirement for the existence of a BIBD is that r and b be integer. Another necessary condition for existence is Fisher's inequality.

Theorem 2.8 (Fisher's inequality; see [51]). *In any BIBD(v,b,r,k,λ), $b \geq v$.*

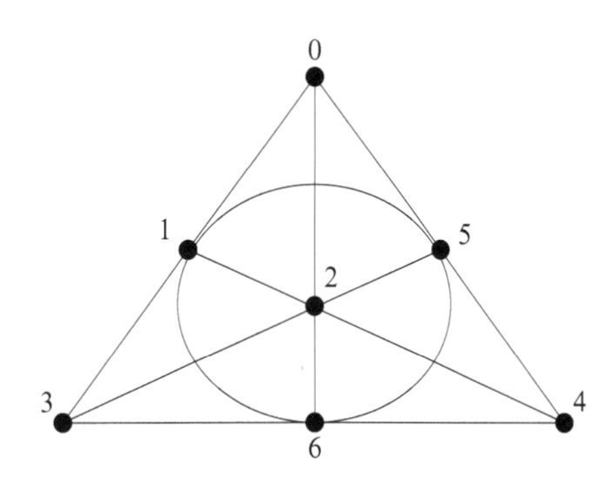

Fig. 2.2 The 2-$(7,3,1)$ design known as the Fano plane

The design most commonly used to illustrate design theoretic concepts is the Fano plane. The Fano plane is a BIBD$(7,3,1)$, also written STS(7). The design has $b = 7$ and $r = 3$. The blocks of the Fano plane are given in Table 2.1, while the Fano plane is represented in Fig. 2.2. Each vertex is a point of the design, and each line (including the centre circle) represents a block of the design.

As we will often be working with triple systems, we note the following two existence theorems.

Theorem 2.9 (See [15]). *There exists an STS(v) if and only if $v \equiv 1,3 \pmod 6$.*

Theorem 2.10 (See [15]). *There exists a TTS(v) if and only if $v \equiv 0,1 \pmod 3$.*

It is worthwhile to note the existence of some other special BIBDs.

Theorem 2.11 (See [51]). *For every prime power $q \geq 2$, there exists a BIBD$(q^2 + q+1, q+1, 1)$.*

The family of BIBDs defined in Theorem 2.11 are known as **projective planes of order q**. Notice that the Fano plane is a projective plane of order two. All projective planes are symmetric BIBDs.

Definition 2.3 (symmetric BIBD). A symmetric BIBD is one in which $b = v$ (or, equivalently, $r = k$).

One of the most important theorems regarding the existence of symmetric BIBDs is the Bruck–Ryser–Chowla theorem.

Theorem 2.12 (Bruck–Ryser–Chowla; see [51]).

1. *If there exists a symmetric BIBD(v,k,λ) with v even, then $k - \lambda$ is a perfect square.*
2. *If there exists a symmetric BIBD(v,k,λ) with v odd, then there exist integers x, y, z (not all zero) such that $x^2 = (k-\lambda)y^2 + (-1)^{(v-1)/2}\lambda z^2$.*

Let H be an Abelian group of order v. A **(v,k,λ)-difference set**, D, is a k-set of elements from H with the property that given any non-zero member $h \in H$, there are precisely λ ordered pairs of elements in D whose difference is h [55]. For $h \in H$, the set $D + h = \{d + h \colon d \in D\}$ is known as a **translate** of D. The design

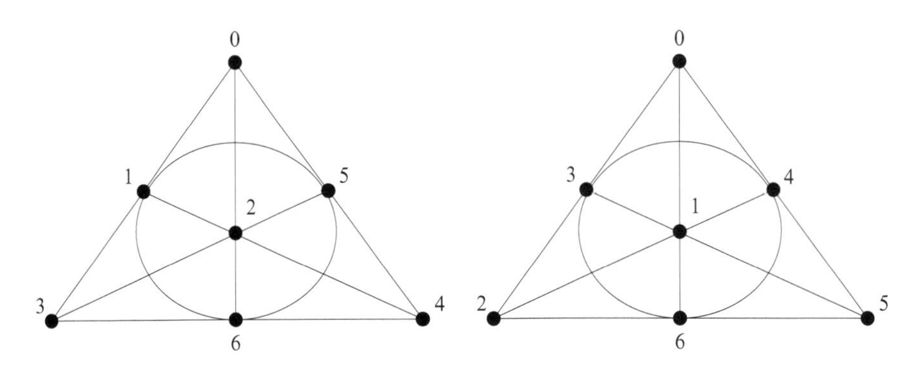

Fig. 2.3 Two isomorphic, but not identical, designs

obtained by developing a difference set D in the group H consists of all translates $\{D+h\colon h \in H\}$. This design is a symmetric BIBD(v,k,λ). An example of a $(7,3,1)$-difference set over $\mathbb{Z}_7$ is $\{1,2,4\}$. The development of difference sets into BIBDs can be generalized as follows. Suppose $D_1,\ldots,D_n$ are k-sets of elements from H. Let λ_{ij} denote the number of ordered pairs of elements in D_j with non-zero difference $h_i \in H$. If all sums $\sum_{j=1}^{n}\lambda_{ij}$ are equal, then developing these sets in H yields a BIBD$(v,nv,nk,k,\sum_{j=1}^{n}\lambda_{1j})$. The sets $D_1,\ldots,D_n$ are called **supplementary difference sets** [55]. For example, a $(9,18,8,4,3)$-supplementary difference set over $\mathbb{Z}_9$ is $\{\{1,2,3,5\},\{1,2,5,7\}\}$. Designs that come from difference sets have special properties. In particular, the relationship of the differences appearing in a difference set determines an automorphism group of the design.

2.2.2 *Automorphisms of BIBDs*

In this section, we look at constructing BIBDs that have prescribed automorphisms. First, we introduce some necessary terminology. An **isomorphism** from a design $(V_1,\mathscr{B}_1)$ to a design $(V_2,\mathscr{B}_2)$ is a bijective map $\phi : V_1 \to V_2$ which maps $\mathscr{B}_1$ to $\mathscr{B}_2$. An **automorphism** of a BIBD is an isomorphism from the BIBD to itself. For example, the two BIBDs shown in Fig. 2.3 are isomorphic but not the same since the first has pair $\{1,3\}$ contained in block $\{0,1,3\}$ whereas the second has pair $\{1,3\}$ contained in block $\{1,3,5\}$. On the other hand, Fig. 2.4 shows the same design twice (although with different pictures—they both have the same set of blocks) and demonstrates that the permutation $(0\ 1)(2\ 5)(3)(4)(6)$ is an automorphism of this BIBD.

The set of all automorphisms of a BIBD forms a permutation group, and any subgroup of the automorphism group is itself a permutation group. Suppose $(V,\mathscr{B})$ is a BIBD and suppose Γ is a subgroup of its automorphism group. The **orbit** of a block $B \in \mathscr{B}$ with respect to Γ is defined

$$\mathrm{orb}_{\Gamma}(B) = \{\pi \cdot B\colon \pi \in \Gamma\},$$

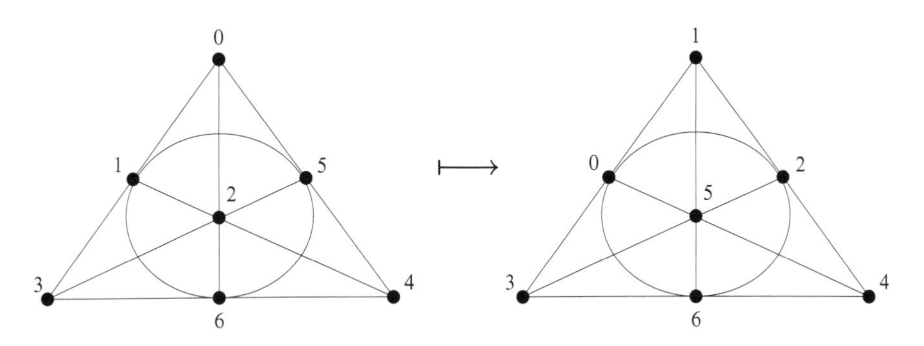

Fig. 2.4 The permutation $(0\ 1)(2\ 5)(3)(4)(6)$ is an automorphism of the Fano plane

where π operates on each element of B independently. For some collection $\mathscr{B}' \subseteq \mathscr{B}$,

$$\mathscr{B} = \cup_{B \in \mathscr{B}'} \{\mathrm{orb}_\Gamma(B)\}.$$

Let $\widehat{\mathscr{B}}$ be a set containing one block from each orbit; in group theoretic terms, this is a set of orbit representatives. We say that $\widehat{\mathscr{B}}$ is a set of **base blocks**, or **starter blocks**, for the BIBD under the action of Γ because $\mathrm{orb}_\Gamma(\widehat{\mathscr{B}}) = \mathscr{B}$. The most common forms of automorphism group used in the construction of designs are those generated by a permutation with a few large cycles and very few fixed points. Overwhelmingly the most common are cyclic groups, $\mathbb{Z}_v, v \in \mathbb{N}$.

Let $V = \mathbb{Z}_v$ be the point set of a BIBD of order v. The design is called **cyclic** if its automorphism group (operating on the points of the system) contains the cyclic group of order v as a subgroup. Without loss of generality, this cyclic group contains the automorphism

$$\pi : i \longmapsto i+1 \pmod{v};$$

therefore, increasing every element of a block by one modulo v results in another block of the design. The Fano plane is a cyclic design, and the first block in Table 2.1 generates the entire design (in fact, any one of the blocks of the design can be used as a base block for the entire design).

It is useful to understand the various behaviours of orbits when working with cyclic designs. While we will later work with cyclic BIBD(v,k,λ)s with general k, here we discuss pairs and triples. The discussion below can easily be extended to base blocks of size k. Let $V = \mathbb{Z}_v$ be the point set of a triple system of order v. All arithmetic will be carried out modulo v. The orbit of a pair $\{x,y\}$, $x,y \in V$, under the action of π (defined above) is $\mathrm{orb}_\pi(\{x,y\}) = \{\{x+i, y+i\} : 0 \leq i \leq v\}$. Orbits of pairs are separated into two classes based on their length. A **full orbit** is an orbit of length v. If v is even, the orbit of $\{x, x \pm v/2\}$ is of length $v/2$; we call this a **half orbit**. We can specify an orbit of pairs uniquely by stating the minimum difference between the two elements. That is, the orbit of the pair $\{x,y\}$ is specified by $\min(|x-y|, |y-x|)$. This is called the **minimum difference** of a pair. Because an orbit of pairs can be specified by a single difference, the most logical set of orbit representatives of pairs is $\{\{0,d\}: 1 \leq d \leq \lfloor v/2 \rfloor\}$. Orbits of the triples in a cyclic triple system can be classified in a similar way. An orbit of triples is of length v unless v is a multiple of three. If v is a multiple of three, the orbit containing $\{0, v/3, 2v/3\}$ has length $v/3$; such an orbit is called a **short orbit**. For each orbit of triples, consider the orbits of pairs associated to it. For example, the triple $\{x,y,z\}$ has associated to it the pair orbits represented by $\min(|x-y|, |y-x|)$, $\min(|y-z|, |z-y|)$ and $\min(|x-z|, |z-x|)$. However, the orbits of pairs associated to a triple do not completely determine the orbit of triples. For example, both triples $\{0,1,3\}$ and $\{0,2,3\}$ in $\mathbb{Z}_7$ are associated with the pair orbits of $\{0,1\}$, $\{0,2\}$ and $\{0,3\}$, but the orbits of these two triples are different.

To represent an orbit of triples, we use an object called a difference triple. Note that a difference triple is not the same as the difference sets defined in the previous section. Given a full orbit of triples, choose an orbit representative $\{0,a,b\}$ such that $a < b$, $a \leq v-b$ and $b-a \leq v-b$. The **difference triple** for this orbit is $(a, b-a, \min(b, v-b))$. In general, a triple (d_1, d_2, d_3) is a difference triple modulo v if and only if $d_i \leq v/2$ and either $d_1+d_2+d_3 \equiv 0 \pmod{v}$ or $d_1+d_2 \equiv d_3 \pmod{v}$. To a short orbit of triples, we associate $\frac{v}{3}$, which is called a **degenerate difference triple**. A **difference partition** of a set is a partition of the set into difference triples.

In Chap. 5, we will work closely with cyclic BIBDs. When discussing cyclic BIBDs, we will refer to the base blocks of the BIBDs. The terminology applied to orbits in the previous paragraph will also be used to describe the properties of base blocks. Suppose $\widehat{\mathscr{B}}$ is a collection of base blocks for a BIBD(v,k,λ). Given a base block $B \in \widehat{\mathscr{B}}$, define $Dev(B) = \{x+B: x \in \mathbb{Z}_v\}$, where $x+B$ denotes the addition of x to every point in B. We call the set Dev(B) the set of blocks **developed** from B. Equivalently, we say that B **generates** the set Dev(B). If B generates v unique blocks, then B is called a **full orbit base block**. If B generates fewer than v unique blocks, then B is called a **short orbit base block**. A base block that generates exactly v/k blocks is called a **regular short orbit base block**. Such a base block must be of the form $\{0, v/k, 2v/k, \ldots, (k-1)v/k\}$. Note that the number of blocks generated by a base block must divide v. By convention, a base block is assumed to include the zero point.

Cyclic triple systems are well studied. The following theorems and lemmas present some relevant existence results.

Lemma 2.1 (See [55]).

1. *Suppose* $v \equiv 1 \pmod{6}$ *and there is a difference partition of* $S = \{1,2, \ldots, (v-1)/2\}$, *then there exists a cyclic STS*(v).
2. *Suppose* $v \equiv 3 \pmod{6}$ *and there is a difference partition of* $S' = \{1,2, \ldots, (v-1)/2\} \setminus \{v/3\}$, *then there exists a cyclic STS*(v).

Proof. Suppose $v \equiv 1 \pmod{6}$ and suppose P is a difference partition of $S = \{1, \ldots, (v-1)/2\}$. Consider a difference triple $T = (x,y,z)$ in P. We will translate T in order to obtain a triple containing zero. Some member of T is congruent to the sum of the other two. Without loss of generality, suppose $z \equiv (x+y) \pmod{v}$.

Table 2.3 An STS(15) constructed from the difference partition $P = \{(1,3,4),(2,6,7)\}$

$\{0,1,4\}$	$\{0,6,13\}$	$\{0,5,10\}$
$\{1,2,5\}$	$\{1,7,14\}$	$\{1,6,11\}$
$\{2,3,6\}$	$\{2,8,0\}$	$\{2,7,12\}$
$\{3,4,7\}$	$\{3,9,1\}$	$\{3,8,13\}$
$\{4,5,8\}$	$\{4,10,2\}$	$\{4,9,14\}$
$\vdots$	$\vdots$	
$\{14,0,3\}$	$\{14,5,12\}$	

Table 2.4 A TS$(7,3)$ constructed from the set of difference triples $\{(1,1,2),(2,2,3),(1,3,3)\}$

$\{0,1,2\}$	$\{0,2,4\}$	$\{0,3,6\}$
$\{1,2,3\}$	$\{1,3,5\}$	$\{1,4,0\}$
$\{2,3,4\}$	$\{2,4,6\}$	$\{2,5,1\}$
$\{3,4,5\}$	$\{3,5,0\}$	$\{3,6,2\}$
$\{4,5,6\}$	$\{4,6,1\}$	$\{4,0,3\}$
$\{5,6,0\}$	$\{5,0,2\}$	$\{5,1,4\}$
$\{6,0,1\}$	$\{6,1,3\}$	$\{6,2,5\}$

We associate $\{0,x,x+y\}$ to T and set $\widehat{\mathscr{B}} = \{\{0,x,x+y\} : (x,y,z) \in P\}$. The triples formed by developing each base block of $\widehat{\mathscr{B}}$ in $\mathbb{Z}_v$ form a Steiner triple system.

Suppose $v \equiv 3 \pmod 6$. The same construction is used; however, we add the blocks generated by $\{0,v/3,2v/3\}$—the base block associated to the degenerate difference triple $(v/3)$. □

To illustrate, we construct an STS(15). Let $S' = \{1,2,3,4,6,7\}$ and $P = \{(1,3,4),(2,6,7)\}$. Then $\widehat{\mathscr{B}} = \{\{0,1,4\},\{0,6,13\},\{0,5,10\}\}$, where $\{0,5,10\}$ is associated to the degenerate difference triple. The STS(15) constructed from the difference partition P is given in Table 2.3.

The construction described in the proof of Lemma 2.1 need not be restricted to the case $\lambda = 1$. For example, suppose we have the difference triples $\{(1,1,2),(2,2,3),(1,3,3)\}$ over the group $\mathbb{Z}_7$. These triples induce a set of base blocks $\widehat{\mathscr{B}} = \{\{0,1,2\},\{0,2,4\},\{0,3,6\}\}$, and these base blocks generate the TS$(7,3)$ presented in Table 2.4. Note that this TS$(7,3)$ has additional automorphisms generated by multiplication by 3 in $\mathbb{Z}_7$.

Theorem 2.13 (See [55]). *There exists a difference partition of* $S = \{1,2,\ldots,(v-1)/2\}$ *for all* $v \equiv 1 \pmod 6$, *and a difference partition of* $S' = \{1,2,\ldots,(v-1)/2\} \setminus \{v/3\}$ *for all* $v \equiv 3 \pmod 6$ *except* $v = 9$.

Corollary 2.1 (See [55]). *There exists a cyclic Steiner triple system of every possible order except nine.*

A complete proof of this theorem can be found in [55]. The proof relies on answers to Heffter's first and second difference problems.

Heffter's first difference problem: Can $\{1,2,\ldots,3m\}$ be partitioned into a collection of m difference triples modulo $6m+1$?

The answer is yes, and this enables us to form an STS$(6m+1)$. For example, if $v = 18s+1$ and $s \geq 2$, the difference partition of S is given by

$$(3r+1, 4s-r+1, 4s+2r+2) \text{ for } r = 0, \ldots, s-1,$$
$$(3r+2, 8s-r, 8s+2r+2) \text{ for } r = 0, \ldots, s-1,$$
$$(3r+3, 6s-2r-1, 6s+r+2) \text{ for } r = 0, \ldots, s-1,$$
$$(3s, 3s+1, 6s+1).$$

Heffter's second difference problem: Can $\{1, 2, \ldots, 3m+1\} \setminus \{2m+1\}$ be partitioned into a collection of m difference triples modulo $6m+3$?
Again, the answer is yes and this enables us to form an STS$(6m+3)$. For example, if $v = 18s+3$ and $s \geq 1$, the difference partition of S' is given by

$$(3r+1, 8s-r+1, 8s+2r+2) \text{ for } r = 0, \ldots, s-1,$$
$$(3r+2, 4s-r, 4s+2r+2) \text{ for } r = 0, \ldots, s-1, v$$
$$(3r+3, 6s-2r-1, 6s+r+2) \text{ for } r = 0, \ldots, s-1.$$

The existence of cyclic triple systems has been completely determined.

Theorem 2.14 (See [15]). *There is a cyclic TS(v, λ) if and only if:*

- $\lambda \equiv 1, 5 \pmod 6$ *and* $v \equiv 1, 3 \pmod 6$, $(v, \lambda) \neq (9, 1)$
- $\lambda \equiv 2, 10 \pmod{12}$ *and* $v \equiv 0, 1, 3, 4, 7, 9 \pmod{12}$, $(v, \lambda) \neq (9, 2)$
- $\lambda \equiv 3 \pmod 6$ *and* $v \equiv 1 \pmod 2$
- $\lambda \equiv 4, 8 \pmod{12}$ *and* $v \equiv 0, 1 \pmod 3$
- $\lambda \equiv 6 \pmod{12}$ *and* $v \equiv 0, 1, 3 \pmod 4$
- $\lambda \equiv 0 \pmod{12}$ *and* $v \geq 3$.

Besides cyclic automorphisms, there are other types of automorphism that may appear in the automorphism group of a BIBD. A BIBD whose automorphism group contains a map which has one fixed point and r cycles of the same length is called ***r*-rotational**. A **1-rotational** BIBD(v, k, λ) is a design that has an automorphism with one fixed point and a cycle of length $v-1$. By convention, the point set of a 1-rotational design of order v is $\mathbb{Z}_{v-1} \cup \{\infty\}$, with ∞ the fixed point, and so the cycle is $(0\ 1 \ldots v-3\ v-2)$. The only 1-rotational designs we will work with are triple systems.

Theorem 2.15 (See [15]). *There is a* 1*-rotational TS(v, λ) if and only if:*

- $\lambda = 1$ *and* $v \equiv 3, 9 \pmod{24}$
- $\lambda \equiv 1, 5 \pmod 6$ *and* $v \equiv 1, 3 \pmod 6$
- $\lambda \equiv 2, 4 \pmod 6$ *and* $v \equiv 0, 1 \pmod 3$
- $\lambda \equiv 3 \pmod 6$ *and* $v \equiv 1 \pmod 2$
- $\lambda \equiv 0 \pmod 6$ *and* $v \neq 2$.

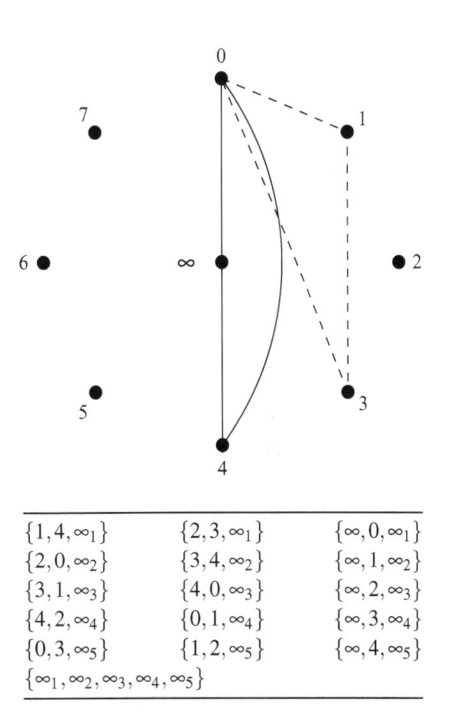

Fig. 2.5 The base blocks of a 1-rotational STS(9) in $\mathbb{Z}_8 \cup \{\infty\}$

Table 2.5 A PBD$(11,\{3,5\},1)$

$\{1,4,\infty_1\}$	$\{2,3,\infty_1\}$	$\{\infty,0,\infty_1\}$
$\{2,0,\infty_2\}$	$\{3,4,\infty_2\}$	$\{\infty,1,\infty_2\}$
$\{3,1,\infty_3\}$	$\{4,0,\infty_3\}$	$\{\infty,2,\infty_3\}$
$\{4,2,\infty_4\}$	$\{0,1,\infty_4\}$	$\{\infty,3,\infty_4\}$
$\{0,3,\infty_5\}$	$\{1,2,\infty_5\}$	$\{\infty,4,\infty_5\}$
$\{\infty_1,\infty_2,\infty_3,\infty_4,\infty_5\}$		

For example, $\{\{\infty,0,4\}\ \{2,3,5\}\}$ is a set of base blocks for a 1-rotational STS(9) in $\mathbb{Z}_8 \cup \{\infty\}$. By convention, the second block should contain a zero; therefore, we replace it with $\{0,1,3\}$. A pictorial representation of these base blocks is given in Fig. 2.5.

2.2.3 *Pairwise Balanced Designs*

The concept of a BIBD can be generalized as follows.

Definition 2.4 (pairwise balanced design). Let v and λ be positive integers and let $K \subseteq \{n \in \mathbb{Z}\colon n \geq 2\}$. A pairwise balanced design, PBD(v,K,λ), is a set system $(V,\mathscr{B})$ with the following properties: (1) $|V| = v$, (2) $|B| \in K$ and $|B| < v$ for all $B \in \mathscr{B}$, and (3) every pair of distinct points is contained in exactly λ blocks.

Note that there need not be blocks of every size in K in a PBD(v,K,λ). Table 2.5 is an example of a PBD$(11,\{3,5\},1)$. The points denoted by ∞ and ∞_i, $1 \leq i \leq 5$, are artefacts of the construction method used. This PBD is 2-rotational with automorphism group generated by the permutation

$$(0\ 1\ 2\ 3\ 4)(\infty_1\ \infty_2\ \infty_3\ \infty_4\ \infty_5)(\infty).$$

Table 2.6 An optimal 2-(6,3,1) packing

$\{0,1,3\}$	$\{1,2,4\}$	$\{0,4,5\}$	$\{2,3,5\}$

Table 2.7 An optimal 2-(6,3,1) covering

$\{0,1,3\}$	$\{0,3,4\}$	$\{0,2,5\}$
$\{1,2,4\}$	$\{1,4,5\}$	$\{2,3,5\}$

2.2.4 *Other Designs*

We will meet several other block designs in this book.

The conditions necessary for the existence of a BIBD preclude their existence for many v when given k, t and λ. In these cases, we study the set systems that come closest to being a BIBD; these are known as packing and covering designs.

Definition 2.5 (t-(v,k,λ) packing). Let t,v,k and λ be positive integers such that $1 \leq t \leq k \leq v$. A t-(v,k,λ) packing is a set system $(V,\mathscr{B})$, such that the following properties hold: (1) $|V| = v$, (2) each block contains exactly k points, and (3) every t-set of V is contained in *at most* λ blocks.

The **packing number**, $\mathrm{D}_\lambda(v,k,t)$, is the maximum number of blocks in any t-(v,k,λ) packing. A packing is said to be **optimal** if $|\mathscr{B}| = \mathrm{D}_\lambda(v,k,t)$. An optimal 2-$(6,3,1)$ packing is given in Table 2.6.

For $v \equiv 1,3 \pmod 6$, an optimal 2-$(v,3,1)$ packing is a Steiner triple system. For $v \equiv 0,2 \pmod 6$ the optimal packing is obtained by deleting a point from an STS$(v+1)$ and all blocks containing that point. For $v \equiv 5 \pmod 6$ the optimal packing is obtained by taking a PBD$(v,\{3,5\},1)$ and replacing the single block of size 5 by two blocks of size 3 intersecting at a point x. Finally, for $v \equiv 4 \pmod 6$, the optimal 2-$(v,3,1)$ packing is obtained from the $v \equiv 5 \pmod 6$ case by deleting one of the points that was on the block of size 5 other than x, and all blocks which contain the deleted point. A general bound is known on the packing number.

Theorem 2.16 (See [52]).

$$D_\lambda(v,k,t) \leq \left\lfloor \frac{v \cdot D_\lambda(v-1,k-1,t-1)}{k} \right\rfloor.$$

Replacing "at most" with "at least" in the definition of packing yields the definition of covering.

Definition 2.6 (t-(v,k,λ) covering). Let t,v,k and λ be positive integers such that $1 \leq t \leq k \leq v$. A t-(v,k,λ) covering is a set system $(V,\mathscr{B})$, such that the following properties hold: (1) $|V| = v$, (2) each block contains exactly k points, and (3) every t-set of V is contained in *at least* λ blocks.

The **covering number**, $\mathrm{C}_\lambda(v,k,t)$, is the minimum number of blocks in any t-(v,k,λ) covering. A covering is said to be **optimal** if $|\mathscr{B}| = \mathrm{C}_\lambda(v,k,t)$. An example of an optimal covering is given in Table 2.7.

Table 2.8 A Latin square of order seven

0	1	2	3	4	5	6
4	5	6	0	1	2	3
1	2	3	4	5	6	0
5	6	0	1	2	3	4
2	3	4	5	6	0	1
6	0	1	2	3	4	5
3	4	5	6	0	1	2

Theorem 2.17 (See [22]).

$$C_\lambda(v,k,t) \geq \left\lceil \frac{v \cdot C_\lambda(v-1,k-1,t-1)}{k} \right\rceil.$$

More information about packing and covering designs can be found in [22, 52]. We turn now to several designs with tabular expressions.

Definition 2.7 (Latin square of order v). A Latin square of order v is a $v \times v$ array in which each cell contains a single symbol from a v-set, V, such that each symbol appears exactly once in each row and exactly once in each column.

A Latin square of order seven is given in Table 2.8.

Definition 2.8 (orthogonal array). An orthogonal array, $\mathrm{OA}(t,k,v)$, is a $v^t \times k$ array with entries from a v-set, V, having the property that in any t columns, each t-tuple of symbols appears exactly once in a row.

For example, an $\mathrm{OA}(3,4,3)$ can be formed by building the 27 rows $(x,y,z,2x+2y+2z)$ for all $x,y,z \in \mathbb{Z}_3$. If L is a Latin square of order v, then an $\mathrm{OA}(2,3,v)$ is formed from the v^2 rows $(i,j,L(i,j))$ for all $1 \leq i,j \leq v$. It is known that for $t=2$, if an $\mathrm{OA}(2,k,v)$ exists, then $k \leq v+1$.

As with BIBDs, when the necessary conditions preclude the existence of a precise object, we can study how close we can get to the object in question.

Definition 2.9 (covering array). A covering array, $\mathrm{CA}(N;t,k,v)$, is an $N \times k$ array filled with entries from a v-set such that, given any set of t columns and any length t vector from the v-set, there exists at least one row where that tuple can be found in those columns.

The **covering array number**, $\mathrm{CAN}(t,k,v)$, is the minimum number of rows in any $\mathrm{CA}(N;t,k,v)$. A covering array is said to be **optimal** if $N = \mathrm{CAN}(t,k,v)$. An optimal covering array is shown in Table 2.9. Note that every pair of columns contains each of the nine possible ordered pairs at least once. For more information on covering arrays, see [12, 13, 26].

Table 2.9 A CA(11;2,5,3)

0	0	0	0	0
0	1	2	2	1
1	0	1	2	2
2	1	0	1	2
2	2	1	0	1
1	2	2	1	0
0	2	1	1	2
2	0	2	1	1
1	2	0	2	1
1	1	2	0	2
2	1	1	2	0

2.2.5 Graphs Related to Block Designs

One natural way to express relationships between blocks of a design is to create a graph having the blocks of the design as vertices. Let $[v]$ denote the set $\{1,2,\ldots,v\}$.

Definition 2.10 (I-block-intersection graph). For $I \subseteq [k]$, the I-block-intersection graph of a BIBD(v,k,λ), $S=(V,\mathscr{B})$, is the graph whose vertex set is $\mathscr{B}$ and whose edge set is $E=\{\{B_1,B_2\}: |B_1 \cap B_2| \in I, B_1, B_2 \in \mathscr{B}\}$.

When $I=[k]$ the I-block-intersection graph is simply called the **block intersection graph**. Denote the I-block-intersection graph of a design $S=(V,\mathscr{B})$ by $G_I^{\mathscr{B}}$, and let $G^{\mathscr{B}}$ represent the block-intersection graph. Figures 2.6 and 2.7 present the 1 block and 2 block-intersection graphs for the TS$(7,3)$ given in Table 2.4. Note that these graphs are on the same vertex set but are edge disjoint. The block-intersection graph for this TS$(7,3)$ is the (edge) union of these two graphs.

Another graphical expression of a block design is the block coloured pair adjacency graph. The **block coloured pair adjacency graph** for a BIBD(v,k,λ) $S=(V,\mathscr{B})$ is the multigraph with vertex set V such that for each pair of distinct vertices a and b, there is an edge between a and b of colour B_i if $\{a,b\} \subseteq B_i \in \mathscr{B}$. Denote the block coloured pair adjacency graph by G^V. Notice that when $\lambda=1$, G^V is simply a complete graph on $|V|$ vertices.

We introduce a variation on G^V for cyclic designs. The **base block coloured pair adjacency graph** for a cyclic design $S=(V,\mathscr{B})$, with base blocks denoted $\widehat{\mathscr{B}}$, is the multigraph with vertex set V such that for each pair of distinct vertices a and b, there is an edge between a and b of colour B_i if the difference $b-a$ is in the base block $B_i \in \widehat{\mathscr{B}}$. Denote the base block coloured pair adjacency graph by G^V_{base}. Given any cyclic design, G^V_{base} will use fewer colours than G^V. An example base block coloured pair adjacency graph is shown in Fig. 2.8. The block coloured pair adjacency graph for the same design would have the same edges, but a larger number of colours.

There are other ways to encode designs in graphs that keep more information about the points.

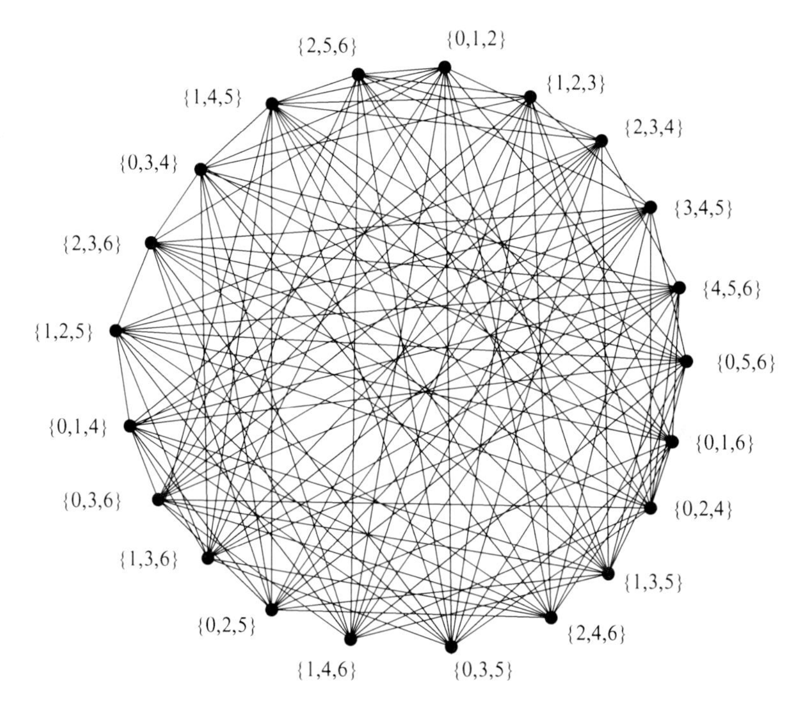

Fig. 2.6 The 1 block-intersection graph of the TS$(7,3)$ given in Table 2.4

Definition 2.11 (point-block incidence graph). The point-block incidence graph of a BIBD(v,k,λ), $S=(V,\mathscr{B})$, is the bipartite graph whose vertex set is $V\cup\mathscr{B}$ and whose edge set is $E=\{\{x,B\}: x\in B\in\mathscr{B}\}$.

The point-block incidence graph of a symmetric BIBD(v,k,λ) is regular of degree k with $|V|=|\mathscr{B}|$. A **$\mathscr{B}$-saturating tour** in the point-block incidence graph is a closed walk that contains each vertex in $\mathscr{B}$ exactly once. Note that vertices from V may appear multiple times in this tour or not at all. This concept will be used in Sect. 5.3.3.

2.2.6 *Configurations*

Recall that a t-(v,k,λ) design is a set system $(V,\mathscr{B})$ such that the following properties hold: (1) $|V|=v$, (2) each block contains exactly k points, and (3) every t-set of distinct points is contained in exactly λ blocks. A **partial design** is a set of elements V and a set of blocks $\mathscr{B}$ such that every t-set occurs in *at most* λ blocks. Configurations are partial designs which, although not stated in the definition, are defined assuming an underlying t-(v,k,λ) design. These are examples of packing designs although they are rarely optimal.

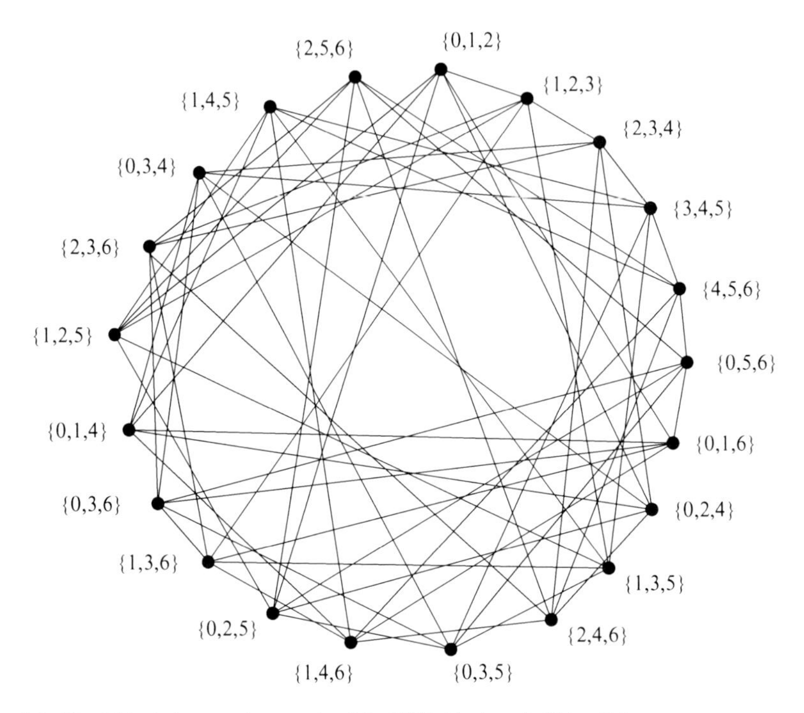

Fig. 2.7 The 2 block-intersection graph of the TS$(7,3)$ given in Table 2.4

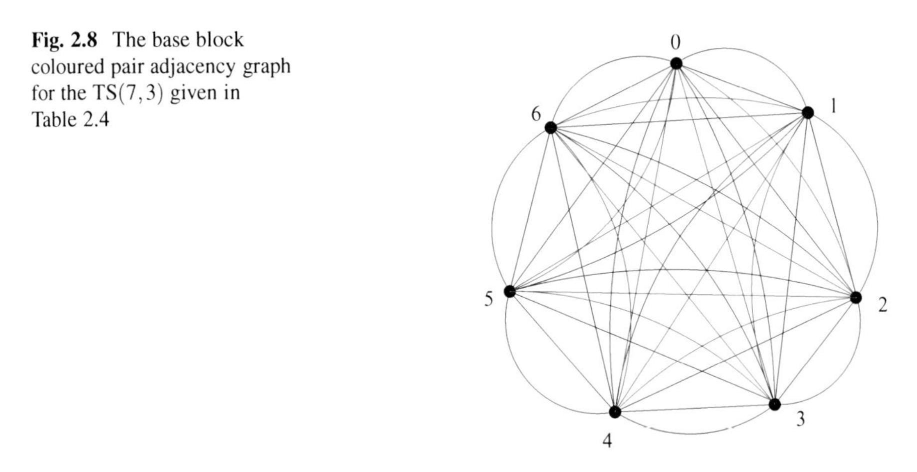

Fig. 2.8 The base block coloured pair adjacency graph for the TS$(7,3)$ given in Table 2.4

Definition 2.12 (*n*-line configuration). A n-line configuration is a collection of n lines (or subsets) having the property that every t-element subset is contained in at most λ lines.

Alternatively, one can specify the number of lines *and* points in a configuration.

Definition 2.13 ((p,ℓ)-configuration). A (p,ℓ)-configuration is a configuration of p points on ℓ lines.

Table 2.10 A $(12,16)$-configuration

$\{1,5,12\}$	$\{1,7,10\}$	$\{2,4,9\}$	$\{1,6,8\}$
$\{3,4,10\}$	$\{3,7,12\}$	$\{4,5,6\}$	$\{5,7,8\}$
$\{2,8,11\}$	$\{1,9,11\}$	$\{2,10,12\}$	$\{2,6,7\}$
$\{3,5,11\}$	$\{6,9,10\}$	$\{4,11,12\}$	$\{3,8,9\}$

Fig. 2.9 The two non-isomorphic $(7,3)$-configurations

The term line is used instead of block to conform with geometric terminology, however, we will use the two terms interchangeably. Table 2.10 is a $(12,16)$-configuration. Note that the point 1 appears with $5,6,7,8,9,10,11,12$ but not with $2,3$ or 4.

Configurations are often presented visually. For example, Fig. 2.9 represents the two non-isomorphic $(7,3)$-configurations. Note that specifying the number of points and blocks of the configuration does not uniquely determine the configuration.

Before we continue, we make a note regarding notation for configurations. All published work regarding configuration ordering, and almost all published work regarding the decomposition of designs into configurations, deals with triple systems; however, these two concepts are easily generalized to BIBDs and PBDs. In this section, as an introduction to configurations, we discuss blocks of size three, while in Chaps. 3, 4 and 5, we will talk about configurations having blocks of size k, $k \in \mathbb{Z}^+$ and in the generalized setting of PBDs. There exist naming conventions for configurations on blocks of size three; thus, given a configuration C on blocks of size three, C' will denote the generalization of this configuration to blocks of size k. When there is some question as to how to generalize a given configuration the generalized version will be explicitly defined.

The study of configurations focuses on the presence or absence of given configurations in triple systems. For example, the number of pairwise non-isomorphic n-line configurations that can occur as blocks of a $\mathrm{TS}(v,\lambda)$ has been determined for $1 \leq n \leq 3$ and arbitrary λ, and for $4 \leq n \leq 8$ when $\lambda = 1$ [23]. The number of configurations of a certain type contained in an $\mathrm{STS}(v)$ has also been studied. For fixed v, non-isomorphic $\mathrm{STS}(v)$s may contain different numbers of the same configuration. Grannell, Griggs and Mendelsohn have determined the number of times each 3-line configuration and the number of times each 4-line configuration appears in an $\mathrm{STS}(v)$ [24], while a general approach for arbitrary designs is given in [1].

Another question often asked regarding configurations is, given a n-line configuration C, can the blocks of an $\mathrm{STS}(v)$ be partitioned into copies of C? If $n|b$ such a decomposition is **exact** [23]. A decomposition that is not exact occurs

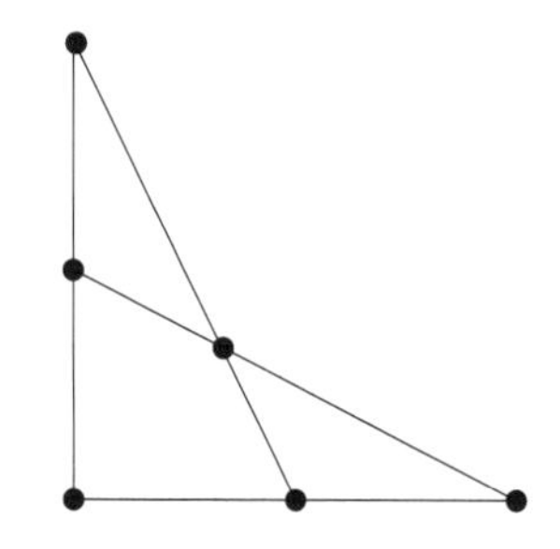

Fig. 2.10 The Pasch configuration

when an STS(v) has an exact decomposition after the deletion of less than n blocks. The seminal paper on decomposition of Steiner triple systems is [28]. The reader interested in decomposition will find the results of Horák and Rosa, and of Mullin, Poplove and Zhu (all contained in [23]) interesting.

The n-line configuration in which every pair of lines intersects in a single common point is called a ***n*-star**. We will use the following result in Chap. 5.

Theorem 2.18 ([28]). *Every STS(v) can be decomposed into 2-stars.*

The 2-star is also denoted A_2. The 2-line configurations we will refer to in this book are shown in Fig. 1.1 (page 4), while the 3-line configurations are shown in Fig. 1.2 (page 5).

Regarding the non-appearance of configurations in designs, questions of the following type are asked: are there triple systems that do not contain a particular type of configuration? Let $B(\lambda)$ be the set of values v for which there exists a TS(v,λ) and let C be a configuration. The **avoidance set** for C and a given λ is defined

$$\Omega(C,\lambda) = \{v\colon\ v \in B(\lambda) \text{ and } \exists \text{ a } \mathrm{TS}(v,\lambda) \text{ not containing } C\}.$$

One of the configurations most often considered in this context is the Pasch, shown in Fig. 2.10. Designs avoiding the Pasch are called **anti-Pasch** or **quadrilateral-free**. The following theorem is the major result in avoidance.

Theorem 2.19 (See [25]). *There exists an anti-Pasch STS(v) for all $v \equiv 1,3 \pmod 6$, except when $v = 7$ or 13.*

Of particular importance to our study, this means that there exist designs that clearly cannot be decomposed into certain configurations.

2.2.7 Decompositions

Balanced incomplete block designs can be thought of as edge disjoint decompositions of complete or complete multipartite graphs into copies of K_k. This idea can easily be generalized to decompositions of any graph into other kinds of subgraph.

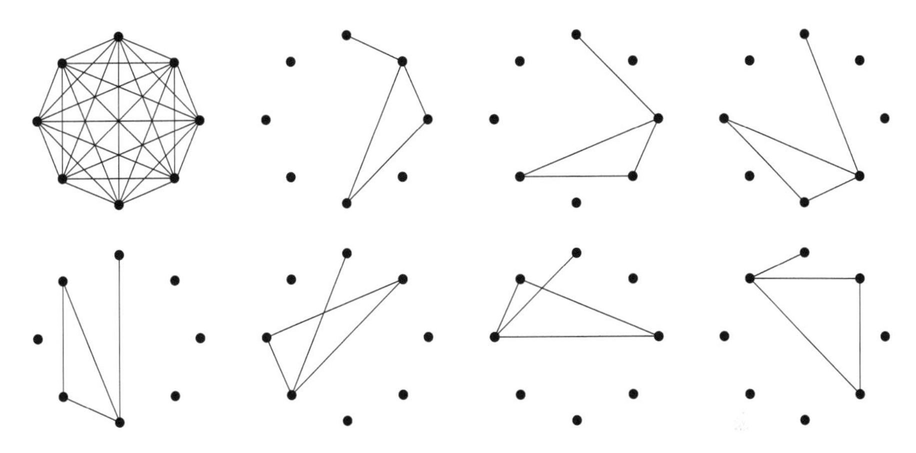

Fig. 2.11 A graph decomposition of K_8 into isomorphic copies of a small graph

Definition 2.14 (*H*-decomposition). Let H and G be graphs. An H-decomposition of G is a set of b edge disjoint subgraphs $H_i \subset G$, $0 \leq i < b$, such that $H_i \cong H$ for all $0 \leq i < b$ and $G = \cup_{i=0}^{b} H_i$.

If H has k vertices and f edges and G has v vertices and e edges, then a necessary condition for such a decomposition to exist is $f|e$ (or $fb = e$, given b in Definition 2.14) and $k \leq v$. Typically, G is a complete graph, a complete multipartite graph or a graph with λ copies of every edge where, as in BIBDs and PBDs, λ is the **index** of the decomposition. The most commonly studied H graphs are K_k, trees (including paths and stars), cubes, closed walks (often incorrectly called "cycles"), cycles, partial matchings (partial 1-factors), and 2-regular spanning graphs [6]. An example of a decomposition of K_8 is given in Fig. 2.11. Configurations can be naturally extended to decompositions by considering how multiple copies of H in the decomposition intersect, in terms of edges and/or vertices. In Sect. 4.3 we will look at orderings of decompositions into matchings, partial matchings and closed walks.

2.3 Combinatorial Orderings

In this section, we introduce the reader to a variety of combinatorial orderings. Research on this topic is wide and deep. Our survey focuses on orderings that provide motivation for ordering the blocks of designs. Note that sequences are denoted by a variety of symbols, including Roman and Greek letters. We retain this inconsistency as it reflects the notation used by the original authors.

2.3.1 *Binary Gray Codes*

A common use of combinatorial algorithms is in the generation of combinatorial structures. The question of how to execute this generation efficiently is of particular importance. One way to make generation efficient is to ensure that successive elements are generated such that they differ in a small, pre-specified way. One of the earliest examples of such a process is Gray code generation. Introduced in 1953 in a pulse code communication system, Gray codes now have applications in areas as diverse as analogue-to-digital conversion, coding theory, switching networks, logic circuits and experimental design (see [44] for an overview). Gray codes have been extensively studied over the past 60 years, and, as a result, there are now many different types of Gray code.

Definition 2.15 (binary Gray code of order *n*). A binary Gray code of order n is a list of all 2^n n-bit strings such that exactly one bit changes from one string to the next.

For the remainder of this section we will refer to binary Gray codes simply as Gray codes. A Gray code is **cyclic** if the change rule also holds from the last n-bit string to the first n-bit string. A binary string of length n can be written as a vector $(a_0, a_1, \dots, a_{n-1})$, where $a_i = 0$ or 1, for $0 \leq i \leq n-1$. The **characteristic vector** for a subset of an n-set is a vector of length n with a 1 in the ith position if i is a member of the subset and 0 if it is not. Therefore, a Gray code can be viewed as a listing of all subsets of a set $X = \{x_0, x_1, \dots, x_{n-1}\}$. It is important to note that we label the characteristic vector positions 1 to n from left to right (opposite to the standard method of labelling bit strings which we will employ in this section). Finally, by converting a binary string to its decimal representation, we obtain a third view of a Gray code: the code is a permutation of the set of the positive integers $\{0, 1, \dots, 2^n - 1\}$.

2.3.1.1 Standard Binary Reflected Gray Cycles

In this section, we look at what might be called the "original" Gray code. The **standard binary reflected cyclic Gray code** is defined as follows. Let Γ_n denote the standard binary reflected cyclic Gray code of order $n \in \mathbb{N}$. Let Γ_0 be the empty list and let $\overline{\Gamma_{n-1}}$ denote the reversal of list Γ_{n-1}, where reversal refers to the order of the list rather than the order of the strings in the list. Let $x\Gamma_{n-1}$, $x \in \mathbb{N}$, indicate that x is prepended to each string in the list Γ_{n-1} and let "," denote concatenation of lists. For $n \geq 1$, the standard binary reflected cyclic Gray code has a simple recursive definition:

$$\Gamma_n := 0\Gamma_{n-1}, 1\overline{\Gamma_{n-1}}.$$

Table 2.11 Standard binary reflected cyclic Gray codes of orders two and three, respectively

Γ_2	Γ_3
00	000
01	001
11	011
10	010
	110
	111
	101
	100

Many Gray codes are created in this recursive manner; write the list of a smaller Gray code or its reversal multiple times, then prepend (or append) a fixed number of bits to each string. Table 2.11 presents two examples of standard binary reflected cyclic Gray codes.

The expression of any binary Gray code can be compressed by specifying a start string (if not specified, assume it is the all zeros string) and then listing the bit position that changes as we move from one string to the next. This list of bit positions is called a **transition sequence**. When a code is cyclic this sequence includes the transition from the last to the first string. We adopt the convention of numbering the bit positions of a string from right to left, labelling the rightmost bit the zero position. The transition sequence for Γ_3, shown in Table 2.11, is 01020102 (we eliminate the commas between bit positions when no confusion will result). Let T_n denote the transition sequence for Γ_n, and let T_n' denote the sequence T_n less the final bit transition. Let T_0 be the empty sequence; then, for $n \geq 1$, the transition sequence for the standard binary reflected cyclic Gray code can also be defined recursively:

$$T_n := T'_{n-1}, n-1, \overline{T'_{n-1}}, n-1 = T'_{n-1}, n-1, T'_{n-1}, n-1. \tag{2.1}$$

The total number of times a bit position appears in a transition sequence is called its **transition count**. When a Gray code is cyclic, every bit position has an even transition count because every bit position must return to its starting value at completion of the cycle. For example, the transition counts for the bit positions of Γ_3 (a cyclic Gray code) are $4, 2$ and 2. Let $d(n)$ denote the number of different transition sequences (assuming a start string of all zeros) that define distinct cyclic Gray codes of order n. A **canonical** transition sequence is one in which the first instance of bit position i appears before the first instance of $i+1$. The transition sequence for Γ_3 in Table 2.11 is a canonical sequence. The only other canonical sequence for a cyclic Gray code of order three is 01210121 (note that this does not produce the standard binary reflected cyclic Gray code). Let $c(n)$ denote the number of canonical transition sequences of order n; then $d(n) = n! \cdot c(n)$ [36]. This is due to the fact that every permutation of the n-bit positions produces another transition sequence. It is interesting to note that $d(2) = 2$, $d(3) = 12$ and $d(4) = 2688$. This is sequence A003042 in [49].

One of the most interesting and useful features of the standard binary reflected cyclic Gray code definition is that it provides an easy way to determine the location of a given bit string within the code without having to determine all strings appearing before it. The location of a given string within a code is called the **rank** of that string. It is also possible to **unrank**; that is, given a rank, we can determine the string in that position of the code. Let

$$\rho : \Gamma_n \to \{0, 1, \ldots, 2^n - 1\}$$

be the function which takes a string in Γ_n and maps it to its rank. Suppose the binary representation of $\rho(\gamma)$, for $\gamma \in \Gamma_n$, is $b_n b_{n-1} \ldots b_1 b_0$. Then $b_n = 0$ and $\rho(\gamma) = \sum_{i=0}^{n} b_i 2^i$. Let γ be represented by $a_{n-1} a_{n-2} \ldots a_0$. Theorem 2.20 defines the relationship between the binary representation of rank and the n-bit string of that rank for the standard binary reflected cyclic Gray code.

Theorem 2.20 (See [38]). *Suppose that $n \geq 1$ is an integer and suppose that $b_n \ldots b_0$ and $a_{n-1} \ldots a_0$ are defined as above. Then, for $j = 0, 1, \ldots, n-1$,*

$$a_j \equiv b_j + b_{j+1} \pmod 2, \tag{2.2}$$

$$b_j \equiv \sum_{i=j}^{n-1} a_i \pmod 2. \tag{2.3}$$

In the next subsection, we discuss binary Gray codes that exhibit additional uniformity properties.

2.3.1.2 Balanced and Monotone Gray Codes

A balanced Gray code is one in which the transition counts are as equal as possible across all bit positions. Note that the standard binary reflected cyclic Gray code of order n is not at all balanced; the rightmost bit changes 2^{n-1} times, whereas the leftmost bit changes only twice (one of these being the return to the beginning of the code).

Definition 2.16 (balanced Gray code of order n). A balanced Gray code of order n is a Gray code for which each transition count is either $\lfloor (2^n - 1)/n \rfloor$ or $\lceil (2^n - 1)/n \rceil$.

Definition 2.17 (balanced cyclic Gray code of order n). Let $a = \lfloor 2^n/n \rfloor$ or $\lfloor 2^n/n \rfloor - 1$, so that a is even. A balanced cyclic Gray code of order n is a Gray code for which each transition count is a or $a + 2$.

A **totally balanced cyclic Gray code** is one in which the transition counts are all equal. Wagner and West [54] have used a graph theoretic argument to show that totally balanced cyclic Gray codes exist when n is a power of two. In [2], Bhat and Savage showed that balanced cyclic Gray codes exist for all $n \in \mathbb{N}$. Table 2.12 shows an example of a balanced cyclic Gray code.

Table 2.12 A balanced cyclic Gray code of order five (read column-wise)

00000	00110	10111	01010
10000	00010	10101	11010
11000	00011	10001	11011
11100	01011	11001	10011
11110	01001	11101	10010
11111	00001	01101	10110
01111	00101	01100	10100
01110	00111	01000	00100

Fig. 2.12 Removing and inserting blocks of strings to improve balance in a Gray code of order three

	000	000	000
	100	001	001
	110	101	011
	010	100	111
	011	110	101
	111	010	100
	101 } flip	011 }	110
	001 }	111 }	010
transition counts:	421	313	232

In 1980, Vickers and Silverman [53] proposed a heuristic to compute balanced Gray codes. The idea is to take any Gray code of order n, remove a block of strings from the end of the code and insert that block between two strings in the remainder of the list, choosing the insertion point so that the resulting list is also a Gray code but with improved balance. While their algorithmic implementation of this idea performed well for $n = 5, 6, 7, 8$, there is no proof as to the efficacy of this approach in general. We give an example (see Fig. 2.12) of the process used by Vickers and Silverman to improve balance in codes because a similar removal and insertion procedure will be employed later in the construction of Ucycles for cyclic BIBDs.

Define the **weight** of a binary string to be the number of one bits it contains. It is not possible to have a Gray code with strings listed by non-decreasing weight, but it is possible to run through weight levels two at a time. Let v_i represent a binary string of length n, and let $w(v_i)$ denote the weight of string v_i.

Definition 2.18 (monotone Gray code of order *n*). Suppose $\Lambda = \ell_0, \ell_1, \ldots, \ldots, \ell_{2^n-1}$ is a Gray code of order n. Λ is monotone if $w(\ell_k) \leq w(\ell_{k+2})$ for $0 \leq k \leq 2^n - 3$. That is, for all $0 \leq i < j < n$, consecutive strings of weight i and $i+1$ precede those of weight j and $j+1$.

Monotone Gray codes have applications to interconnection networks and also provide some insight into the middle levels problem [45]. Savage and Winkler have proved, by an elegant construction, that there exists a monotone Gray code for each order $n \in \mathbb{N}$ [45]. Table 2.13 shows a monotone Gray code of order five.

Table 2.13 A monotone Gray code of order five (read column-wise)

00000	11000	01010	11110
00001	10000	01011	11100
00011	10001	01001	11101
00010	10101	01101	11001
00110	10100	00101	11011
00100	10110	00111	10011
01100	10010	01111	10111
01000	11010	01110	11111

2.3.2 *Combinatorial Gray Codes*

The notion of a Gray code can be extended to combinatorial objects other than binary strings. Such listings have come to be known as **combinatorial Gray codes**. In this section, we look at Gray codes for k-subsets of n-sets, Gray codes for permutations of n-sets and Gray codes for vectors.

2.3.2.1 *k*-subsets of *n*-sets

There are several ways to represent k-subsets of an n-set. For example, if we are considering 4-subsets of $\{1,2,\ldots,10\}$, we have the following representations of the subset $\{2,5,7,4\}$: (1) unordered: $\{2,5,7,4\}$, (2) ordered: $(2,4,5,7)$, and (3) characteristic vector: $(0,1,0,1,1,0,1,0,0,0)$.

We are interested in listing all k-subsets of an n-set such that consecutive subsets differ in a small way. One such listing is a minimal change ordering. A list is in **minimal change order** if consecutive subsets differ by the deletion of one element and the insertion of another. In Sect. 2.3.1 we saw that there exists a binary Gray code of order n, for all $n \in \mathbb{N}$. Considering the n-bit strings of a binary Gray code to be characteristic vectors, we have a minimal change listing for all subsets of $[n] = \{1,2,\ldots,n\}$ (here the deletion or the insertion, but not both, may be the null element). Therefore, from a binary Gray code of order n, we can obtain a listing of all k-subsets of $[n]$, represented by characteristic vectors, simply by deleting all vectors except those of weight k. Denote such a list, obtained from the standard binary reflected cyclic Gray code, by $\Gamma_{n,k}$. The following result is due to Bitner et al.

Theorem 2.21 ([3]). *Successive binary vectors in $\Gamma_{n,k}$ differ in exactly two bit positions.*

While this is a surprising and nice feature of the standard binary reflected cyclic Gray code, in most cases we would rather not generate a complete cyclic binary Gray code of order n and then delete most strings to obtain the Gray code for k-subsets of an n-set. Luckily there are several methods for directly generating minimal change orderings for all k-subsets of an n-set. Denote the set of $\binom{n}{k}$ k-subsets of $[n]$ by $\mathscr{S}_k^n$. As with binary Gray codes, once we have established that

Table 2.14 Revolving door lists for 3-subsets of [5] and 4-subsets of [5], respectively

$A_{3,5}$	$A_{4,5}$
$\{1,2,3\}$	$\{1,2,3,4\}$
$\{1,3,4\}$	$\{1,2,4,5\}$
$\{2,3,4\}$	$\{2,3,4,5\}$
$\{1,2,4\}$	$\{1,3,4,5\}$
$\{1,4,5\}$	$\{1,2,3,5\}$
$\{2,4,5\}$	
$\{3,4,5\}$	
$\{1,3,5\}$	
$\{2,3,5\}$	
$\{1,2,5\}$	

the subsets of $\mathscr{S}_k^n$ can be listed in a minimal change order, additional constraints will be added to the definition of minimal change so that the elements allowed to change from one subset to the next are restricted.

Revolving Door

The most common method for generating minimal change listings for the k-subsets of $[n]$ is the revolving door algorithm. The revolving door ordering is based on Pascal's identity:

$$\binom{n}{k} = \binom{n-1}{k-1} + \binom{n-1}{k}.$$

In the language of k-subsets of $[n]$, this identity represents the fact that the collection of k-subsets of an n-set can be partitioned into two disjoint sub-collections: the subsets that contain n and the subsets that do not contain n. Let $A_{n,k}$ denote the revolving door ordering for $\mathscr{S}_k^n$. For $k = 0$, $A_{n,0} = \emptyset$ and for $k = n$, $A_{n,n} = \{1,2,\ldots,n\}$. Let $A_{n-1,k-1} \cup \{n\}$ denote the list $A_{n-1,k-1}$ with the element n added to each set and let $\overline{A_{n,k}}$ denote the list $A_{n,k}$ in reverse order. For $0 < k < n$,

$$A_{n,k} := A_{n-1,k}, \overline{A_{n-1,k-1}} \cup \{n\}.$$

Theorem 2.22 (See [38]). *For any integers k and n such that $1 \leq k \leq n$, the list $A_{n,k}$ is a minimal change listing for $\mathscr{S}_k^n$.*

Table 2.14 gives two examples of minimal change listings for k-subsets of [5] obtained using the revolving door algorithm.

Strong Minimal Change

The following property is known to Wilf as strong revolving door [57] and to others as strong minimal change [20]. A **strong minimal change listing** for k-subsets of

Table 2.15 A strong minimal change list for 3-subsets of [6] (read column-wise)

(1,2,3)	(1,4,6)	(2,3,4)	(2,4,5)
(1,2,4)	(1,4,5)	(2,3,5)	(3,4,5)
(1,2,5)	(1,3,5)	(2,3,6)	(3,4,6)
(1,2,6)	(1,3,6)	(2,5,6)	(3,5,6)
(1,5,6)	(1,3,4)	(2,4,6)	(4,5,6)

$[n]$ is a minimal change listing with the additional restriction that two consecutive subsets, when written in sorted order, differ only in a single component. That is, if the ith element of an ordered set leaves, then the new element must also fall in the ith position of the new ordered set. Eades and McKay [20] give a recursive definition for strong minimal change lists. Let $L_{n,k}$ denote a listing of $\mathscr{S}_k^n$ for which the strong minimal change condition is satisfied. For $k = 0$, $L_{n,0} = \emptyset$, for $k = 1$, $L_{n,1} = (1), (2), \ldots, (n)$, and for $k = n$, $L_{n,n} = (1, 2, \ldots, n)$. Let $L_{n,k} + x$ denote the list where $x \in \mathbb{N}$ has been added to each entry of each ordered subset in the list $L_{n,k}$ and let $(x, L_{n,k})$ denote the list of ordered subsets $L_{n,k}$ with x joined as the first entry of each ordered subset. For $1 < k < n$,

$$L_{n,k} := (1, 2, L_{n-2,k-2} + 2), (1, \overline{L_{n-2,k-1}} + 2), (L_{n-1,k} + 1).$$

Table 2.15 gives an example of a strong minimal change listing.

Adjacent Change

An even stronger restriction on the listing of ordered k-subsets of $[n]$ is to require that consecutive ordered subsets differ only in a single component *and* that the change be an increase or decrease by one. This property is known as **adjacent change** due to the fact that if the k-subsets are in characteristic vector representation, this restriction requires that consecutive vectors be of the form $(\ldots, x_i, x_{i+1}, 1, 0, x_{i+4}, \ldots)$ and $(\ldots, x_i, x_{i+1}, 0, 1, x_{i+4}, \ldots)$. This change may be referred to as a transposition of bits in the binary vectors.

The adjacent change requirement is so restrictive that it is not always possible to find a list satisfying the property. Eades et al. [19], and independently Buck and Wiedemann [7], have shown that such a listing exists only if $k = 0, 1, n-1, n$ or if n is even and k is odd. Ruskey proves this result in [40] by presenting a construction algorithm.

Since a strong minimal change listing is always possible but an adjacent change listing is not, it is reasonable to look for a restriction that falls between these two and for which it is always possible to construct a listing. Independently, Chase [9] and Ruskey [41] have proved that it is always possible to create a list with a relaxed form of the adjacent change property. Let two distinct binary vectors be **two-close** if they differ by a transposition of two bits that are either adjacent or have a single zero between them. In [41], Ruskey provides a recursive construction for such a listing.

Table 2.16 Trotter–Johnson listings for permutations of orders one, two and three, respectively

J_1	J_2	J_3
1	1 **2**	1 2 **3**
	2 1	1 **3** 2
		3 1 2
		3 2 1
		2 **3** 1
		2 1 **3**

2.3.2.2 Permutations

In this section, we are interested in listing all $n!$ permutations of an n-set such that consecutive permutations differ in a small way. A **minimal change listing of permutations** is a list in which consecutive permutations differ in exactly two positions. That is, one permutation can be obtained from the other by a single transposition of elements. It is always possible to create such a list [48], so it is natural to restrict the type of transpositions allowed and ask if it is possible to list all permutations such that consecutive permutations differ by an allowed transposition.

The most common method for generating a minimal change listing of permutations is the Trotter–Johnson algorithm. The list generated by this algorithm has the property that the positions in which consecutive permutations differ are adjacent. This is equivalent to saying that the set of allowed transpositions is $\{(1\ 2),(2\ 3),\ldots,(i\ (i+1)),\ldots,((n-1)\ n)\}$. It is important to recognize that the transpositions act on the positions of an n-tuple. That is, $(i\ i+1)$ swaps the symbols in *positions* i and $i+1$ in a given n-tuple; it does not swap the *values* i and $i+1$. As with many combinatorial Gray code constructions, the Trotter–Johnson algorithm is recursive. Suppose $J_{n-1} = \pi_0, \pi_1, \ldots, \pi_{(n-1)!-1}$ is a minimal change listing for the $(n-1)!$ permutations of $[n-1]$. Form a new list by repeating each permutation in the list n times, that is, $\pi_0, \ldots, \pi_0, \pi_1, \ldots, \pi_1, \ldots, \pi_{(n-1)!-1}, \ldots, \pi_{(n-1)!-1}$. Now insert element n into each of the n copies of π_i as follows. If i is even, first place n after the element in position $n-1$, then after the element in position $n-2$, and so on until n is inserted at the beginning of the last π_i. If i is odd, proceed with insertion in the opposite order. The resulting list is J_n. Table 2.16 shows three Trotter–Johnson listings. The value n is in bold to illustrate the pattern of insertion. The convention for these lists is to write permutations without brackets.

The Trotter–Johnson construction shows that it is possible to list all permutations of $[n]$ such that each permutation can be obtained from the previous by the transposition of adjacent elements. Kompel′maher and Liskovec [37], and independently Slater [48], have shown that it is possible to list all permutations of $[n]$ such that each permutation can be obtained from the previous by a single transposition from an arbitrary fixed basis of transpositions for S_n—the symmetric group on n elements. The proof of Kompel′maher and Liskovec is technical and involves building a transition sequence of transpositions for $n+1$ from a sequence of transpositions for n. Slater takes a graph theoretical approach to the question which simplifies the proof.

Table 2.17 A monotone adjacent change Gray code for $\mathscr{V}_{3+}^{3}$ (read column-wise)

(0,0,0)	(2,0,0)	(1,0,2)	(0,0,3)	(1,3,3)	(2,3,0)	(3,2,2)
(0,0,1)	(2,0,1)	(1,1,2)	(0,1,3)	(1,2,3)	(3,3,0)	(3,2,1)
(0,1,1)	(2,1,1)	(1,2,2)	(0,2,3)	(1,1,3)	(3,3,1)	(3,2,0)
(0,1,0)	(2,1,0)	(1,2,1)	(0,3,3)	(1,0,3)	(3,3,2)	(3,1,0)
(1,1,0)	(2,2,0)	(1,2,0)	(0,3,2)	(2,0,3)	(3,3,3)	(3,1,1)
(1,1,1)	(2,2,1)	(0,2,0)	(0,3,1)	(2,1,3)	(3,2,3)	(3,0,1)
(1,0,1)	(2,2,2)	(0,2,1)	(0,3,0)	(2,2,3)	(3,1,3)	(3,0,0)
(1,0,0)	(2,1,2)	(0,2,2)	(1,3,0)	(2,3,3)	(3,0,3)	
	(2,0,2)	(0,1,2)	(1,3,1)	(2,3,2)	(3,0,2)	
		(0,0,2)	(1,3,2)	(2,3,1)	(3,1,2)	

2.3.2.3 Vectors

Recent work of Dewar, McInnes and Proos [18] deals with Gray codes for vectors. Let $\mathscr{V}_{\mathbf{n}}^{k}$ denote the set of k-dimensional vectors with integer coordinates in $[-n,n] \subset \mathbb{N}$ and let $\mathscr{V}_{\mathbf{n}+}^{k}$ denote the set of k-dimensional vectors with integer coordinates in $[0,n] \subset \mathbb{N}$. They say a collection of vectors is in **adjacent change order** if consecutive vectors differ in exactly one coordinate and the difference is ± 1. They go on to define **monotone adjacent change Gray codes for a set of vectors** $\mathscr{V}$ as an adjacent change Gray code with the vectors in strictly non-decreasing L_∞ norm order. This terminology references Savage and Winkler's definition of monotone Gray codes for binary strings [45]. Dewar, McInnes and Proos have shown that there exist monotone adjacent change Gray codes for $\mathscr{V}_{\mathbf{n}+}^{k}$, for all $n,k \in \mathbb{N}$. Further, there exist monotone adjacent change Gray codes for $\mathscr{V}_{\mathbf{n}}^{k}$, for all $n,k \in \mathbb{N}$, if and only if k is even. Their construction employs methodology similar to that of the recursive constructions described earlier in this chapter. An example is given in Table 2.17.

2.3.3 De Bruijn Sequences

De Bruijn sequences are a type of minimal change listing in which successive strings differ not in a single bit position but in a small structural way. A de Bruijn sequence of order n is a listing of all n-bit stings such that successive strings differ by a shift one position left (dropping the first element) followed by the introduction of a new last element. This allows us to compress the list of all n-bit strings into a single cyclic list of 2^n bits.

Definition 2.19 (de Bruijn sequence of order *n*). A de Bruijn sequence of order n is a circular binary sequence of length 2^n in which every n-bit string appears as a contiguous subsequence.

Figure 2.13 represents a de Bruijn sequence of order four. De Bruijn sequences exist for all positive integers n. The standard proof involves first defining a de Bruijn digraph, then showing that such a graph has an Euler tour. Recall that an Euler tour

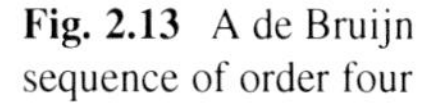

Fig. 2.13 A de Bruijn sequence of order four

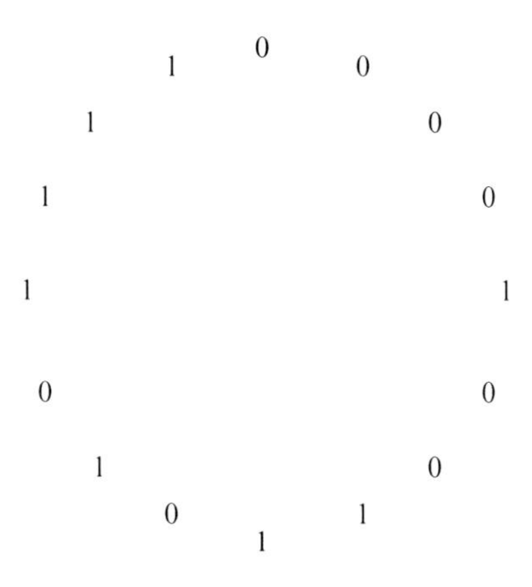

is a tour that uses each edge of a graph exactly once. Define the **de Bruijn digraph of order *n***, denoted $\mathscr{G}_n$, as follows. Let the vertices of the digraph be the $(n-1)$-bit strings. Join vertex $v = x_0x_1 \dots x_{n-2}$ by directed edges to vertices $v_0 = x_1 \dots x_{n-2}0$ and $v_1 = x_1 \dots x_{n-2}1$. Note that there are exactly 2^n edges in this digraph and that there exists an edge corresponding to each n-bit string—the string $x_0x_1 \dots x_{n-1}$ corresponds to the edge $(x_0x_1 \dots x_{n-2}, x_1 \dots x_{n-2}x_{n-1})$. A de Bruijn sequence of order n is equivalent to an Euler tour in $\mathscr{G}_n$. Figure 2.14 shows the de Bruijn digraph of order four.

The de Bruijn digraph of order n is a strongly connected 2-in 2-out graph and is therefore Eulerian (see Theorem 2.5). Consequently, a de Bruijn sequence of order n exists for all positive integers n. An interesting method for constructing de Bruijn sequences is due to Martin [39]. Given a set of r different symbols, consider the r^n arrangements of n symbols with repetitions of the symbols allowed. Can a sequence of these symbols be constructed such that each of these r^n arrangements is found exactly once as a subsequence of n consecutive symbols in this sequence? Martin's algorithm shown in Algorithm 2.1 proves, by construction, that this question can be answered to the affirmative.

For the binary alphabet, this algorithm produces a de Bruijn sequence. In this case, the construction method amounts to applying the rule: start with $n-1$ zeros, then add a one as long as the new n-bit string is not a repeat of one already constructed; otherwise, add a zero.

De Bruijn sequences are named after N. de Bruijn who published a paper on their existence in 1946. However, these sequences were first discovered, and their existence proved by Flye-Sainte Marie in 1894. De Bruijn sequences are also known as **full length nonlinear shift register cycles**. Various algorithms for determining at least one or all possible de Bruijn sequences for a given n are presented in [21].

It is possible to generate a de Bruijn sequence of order n from a de Bruijn sequence of order $n-1$. This idea is reminiscent of the recursive Gray code constructions discussed in Sects. 2.3.1 and 2.3.2. Define an onto map between the set of 2^n binary strings of length n, denoted by V_n, and the set of 2^{n-1} binary strings of length $n-1$, denoted by V_{n-1}, as follows:

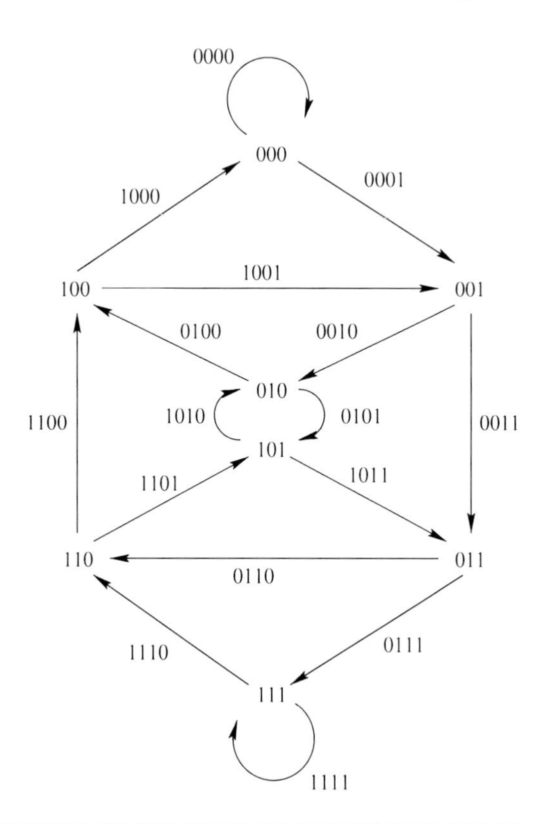

Fig. 2.14 The de Bruijn digraph of order four

Algorithm 2.1 Martin's algorithm for constructing de Bruijn sequences

Denote the r different symbols $e_0, e_1, \ldots, e_{r-1}$.
Set each of the first $n-1$ symbols of the constructed sequence S to e_0.
repeat
  Suppose $S = a_0a_1 \ldots a_{n-2}a_{n-1} \ldots a_{m-n} \ldots a_{m-2}$
  Let e_i be the symbol with the greatest subscript such that $a_{m-n} \ldots a_{m-2}a_{m-1}$ has not yet
    appeared in S.
  Append symbol e_i to S as a_{m-1}.
until no symbol can be appended to S

$$f : V_n \longrightarrow V_{n-1}$$
$$(\mathbf{a} = a_0a_1 \ldots a_{n-1}) \longrightarrow (\mathbf{b} = b_0b_1 \ldots b_{n-2}),$$

where $f(\mathbf{a}) = \mathbf{b}$ if and only if $b_i = a_i + a_{i+1}$ for $i = 0, 1, \ldots, n-2$ [21]. This relation implies that given an $(n-1)$-tuple $\mathbf{b}$ in V_{n-1} and a single bit of $\mathbf{a}$, say, a_0, the next $n-1$ bits of $\mathbf{a}$ can be uniquely determined using the equation $a_{i+1} = a_i + b_i$, for $i = 0, 1, \ldots, n-2$, or equivalently, $a_{i+1} = a_0 + b_0 + b_1 + \cdots + b_i$.

Let C be a de Bruijn sequence of order $n-1$. In this paragraph we abuse the definition of f by discussing the application of f to sequences of length 2^n rather than strings of length n. That is, if $t_0, t_1, \ldots, t_{2^n-1}$ is the image under f of

$s_0,s_1,\ldots,s_{2^n-1}$, then $t_i = s_i + s_{i+1}$, for $i = 0,1,\ldots,2^n-1$, with subscript addition modulo 2^n. C is the image under f of two sequences in V_n, obtained by choosing $a_0 = 0$ or $a_0 = 1$. Denote the pre-image of C having $a_0 = 0$ by C_0 and denote the pre-image of C having $a_0 = 1$ by C_1. Clearly the bitwise sum modulo 2 of C_0 and C_1 is the all ones string. We can join C_0 and C_1 at a common $(n-1)$-bit string to create a de Bruijn sequence of order n. The two sequences must share at least one common $(n-1)$-bit string. To see this, consider the all ones string of length $(n-1)$ in C. The pre-image of this string under f is either the n-bit string of alternating zeros and ones starting with 0 or the n-bit string of alternating zeros and ones starting with 1; each of these strings appears in one of C_0 and C_1; therefore, the $(n-1)$-bit string of alternating zeros and ones starting with 0 appears in both sequences. Note that we can think of each C_i sequence as a cycle. Each of the 2^{n-1} strings of length n appearing in C_i, for $i \in \{0,1\}$, is distinct, and each sequence contains a distinct set of strings (otherwise C would contain a repeated $(n-1)$-bit string). To create a de Bruijn sequence of order n, splice one cycle into the other by writing the first cycle up to and including a common $(n-1)$-tuple, then writing the second cycle starting from the common $(n-1)$-tuple and cycling around to write the full cycle, finishing again with the common $(n-1)$-tuple, and finally completing the first cycle. For example, $C = 00011101$ is a de Bruijn sequence of order three with pre-images $C_0 = 00001011$ and $C_1 = 11110100$. We have a choice of where to splice the two cycles together as they have several length three subsequences in common. Arbitrarily choose the common sequence 101, which yields 0000101**00111101**1—a de Bruijn sequence of order four.

2.3.4 Universal Cycles

Universal cycles are a generalization of de Bruijn sequences to families of combinatorial objects other than binary strings. The definition was proposed by Chung et al. in 1992 [10].

Definition 2.20 (universal cycle). Let $\mathscr{F}_n$ be a family of combinatorial objects of "rank" n and let $m = |\mathscr{F}_n|$. Assume each $F \in \mathscr{F}_n$ is specified by some sequence $x_0,x_1,\ldots,x_{n-1}$, where, for $0 \le i \le n-1$, $x_i \in A$ for some fixed alphabet A. $\mathbf{U} = a_0,a_1,\ldots,a_{m-1}$ is a universal cycle (or Ucycle) for $\mathscr{F}_n$ if $a_{i+1},\ldots,a_{i+n}$, $0 \le i < m$, runs through each element of $\mathscr{F}_n$ exactly once, where index addition is performed modulo m.

Note that the term rank does not refer to the size of each combinatorial object, but to the size of its representative. As an illustrative example, a de Bruijn sequence of order n is a Ucycle for the family of binary strings of length n, so $\mathscr{F}_n = \{x_0\ldots x_{n-1} : x_i \in \{0,1\}, 0 \le i \le n-1\}$ and $m = 2^n$. In this case, each binary vector represents itself; however, for other families, we will see that the representative of each object may be different from the object and may not be the same size as the

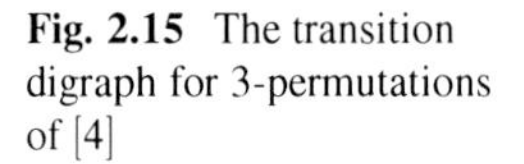
Fig. 2.15 The transition digraph for 3-permutations of [4]

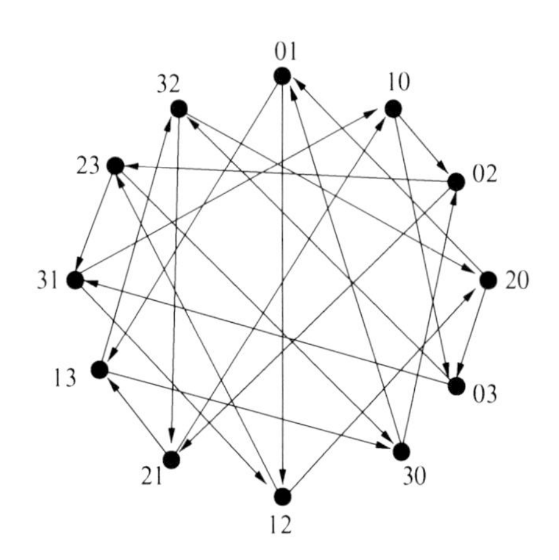

object. The next three sections review the literature on the existence of Ucycles for permutations, subsets and partitions.

2.3.4.1 Permutations

We divide our discussion of Ucycles for permutations into two parts: (1) k-permutations of an n-set where $k < n$ and (2) n-permutations of an n-set. By the set of k-permutations of an n-set, we mean all $k!\binom{n}{k}$ arrangements of k symbols from $[n]$, not permutations in the sense of the symmetric group. Thus, 12 and 21 are distinct elements, and we write these elements without brackets to avoid confusion.

The existence of Ucycles for k-permutations of $[n]$ was proved by Jackson [34]. The proof is relatively easy (as compared with Ucycles for other combinatorial objects) as it employs a transition digraph. We have seen transition digraphs already—the de Bruijn digraph is an example of one. In general, a **transition digraph** for a family of combinatorial objects of order n is a digraph where vertices represent the objects of order $n-1$ and where each edge represents one of the objects of order n. Define the transition digraph for k-permutations of $[n]$, denoted $T_{n,k}$, as follows. The vertices are $(k-1)$-permutations of $[n]$. A directed edge (u,v) exists if the $k-2$ length tail of vertex u is the same as the $k-2$ length head of vertex v and the head element of vertex u is not equal to the tail element of vertex v, that is, $u = x_0x_1 \ldots x_{k-2}$ and $v = x_1 \ldots x_{k-2}x_{k-1}$, $x_0 \neq x_{k-1}$. The edge (u,v) represents the permutation $(x_0\ x_1\ \ldots x_{k-1})$, which is why it is necessary that x_0 be different from x_{k-1}. Figure 2.15 shows the transition digraph for 3-permutations of [4]. Notice that the edge $(10,02)$ exists but the edge $(02,20)$ does not.

Theorem 2.23 ([34]). *For every integer $k \geq 3$ and every integer $n \geq k+1$, there exists a Ucycle for the k-permutations of $[n]$.*

We present the proof in full to illustrate the methods often employed in establishing the existence of Ucycles and to clarify some of Jackson's statements.

We will require the following lemma which is stated but not proved by Jackson. For completeness, we present the proof here. Define a **cyclic rotation** of a t-permutation to be a shift of $s < t$ contiguous elements from the front of the permutation to the end. That is, $i_s i_{s+1} \dots i_{t-2} i_{t-1} i_0 i_1 \dots i_{s-1}$ is a cyclic rotation of $i_0 i_1 \dots i_{t-2} i_{t-1}$.

Lemma 2.2. *Let $v = v_0 v_1 \dots v_{k-2}$ denote a vertex of $T_{n,k}$. There is a directed path in $T_{n,k}$ from v to any cyclic rotation of v.*

Proof. Suppose we are interested in reaching the cyclic rotation $v' = v_{m-1} \dots v_{k-3} v_{k-2} v_0 \dots v_{m-2}$, $1 < m < k$. We will prove that there exists a directed path from v to v' in the transition digraph. Given a set S and a subset T, let $\overline{T} = S \setminus T$. The following path certainly exists in $T_{n,k}$:

$$v_0 v_1 \dots v_{k-2} \longrightarrow v_1 \dots v_{k-2} b_0 \longrightarrow$$
$$v_2 \dots v_{k-2} b_0 b_1 \longrightarrow \cdots \longrightarrow v_{m-1} \dots v_{k-2} b_0 \dots b_{m-2},$$

where $b_j \in \overline{\{v_j, v_{j+1}, \dots, v_{k-2}, b_0, \dots, b_{j-1}\}}$, for all $0 \le j \le m-2$. The set $\{v_j, v_{j+1}, \dots, v_{k-2}, b_0, \dots, b_{j-1}\}$ contains $k-1$ elements; therefore, it is always possible to select an appropriate b_j from $[n]$. Continuing the established path, the vertex $v_{m-1} \dots v_{k-2} b_0 \dots b_{m-2}$ is adjacent to $v_m \dots v_{k-2} b_0 \dots b_{m-2} b_{m-1}$. It is now possible to begin moving towards the permutation v'. The value denoted by v_{m-1} can be added to the right end of $v_m \dots v_{k-2} b_0 \dots b_{m-2}\, b_{m-1}$, to obtain a valid k-permutation, since $v_{m-1} \in \overline{\{v_m, \dots, v_{k-2}, b_0, \dots, b_{m-1}\}}$; hence, $v_m \dots v_{k-2} b_0 \dots b_{m-2} b_{m-1}$ is adjacent to $v_{m+1} \dots v_{k-2} b_0 \dots b_{m-2} b_{m-1} v_{m-1}$ in $T_{n,k}$. Continue the path by adding elements on the right in the order desired for the final permutation. The following path can be created:

$$v_{m+1} v_{m+2} \dots v_{k-2} b_0 \dots b_{m-2} b_{m-1} v_{m-1} \to v_{m+2} \dots v_{k-2} b_0 \dots b_{m-2} b_{m-1} v_{m-1} v_m \to \cdots$$
$$\cdots \to b_0 b_1 \dots b_{m-1} v_{m-1} \dots v_{k-2} \to b_1 \dots b_{m-1} v_{m-1} \dots v_{k-2} v_0 \to \cdots$$
$$\cdots \to v_{m-1} v_m \dots v_{k-2} v_0 \dots v_{m-2}.$$

Thus, there exists a directed path from v to any cyclic rotation of v. □

Proof (Proof of Theorem 2.23). The existence of a Ucycle for k-permutations of $[n]$ is equivalent to the existence of an Euler tour in the transition digraph, $T_{n,k}$. To prove an Euler tour exists, we show that $T_{n,k}$ is strongly connected and that each vertex of $T_{n,k}$ has in-degree equal to out-degree (Theorem 2.5).

To show that $T_{n,k}$ is strongly connected, it is sufficient to show that v is connected to $12\dots(k-1)$ and that $12\dots(k-1)$ is connected to v, for any vertex v. Let $v = v_0 v_1 \dots v_{k-2}$ be a vertex of $T_{n,k}$. By connected, we mean that there exists a directed path from the first vertex to the second. This is done in two steps:

1. Show v is connected to some $w = w_0 w_1 \dots w_{k-2}$, where $w_0 < w_1 < \cdots < w_{k-2}$, then
2. Show w is connected to $12\dots(k-1)$.

If $v_0 < v_1 < \cdots < v_{k-2}$, the first step is complete. Otherwise, $v_m < v_{m-1}$ for some $m \in \{1,2,\ldots,k-2\}$. Lemma 2.2 implies that there exists a directed path from $v_0v_1 \ldots v_{m-1}v_m \ldots v_{k-2}$ to $v_{m-1}v_m \ldots v_{k-2}v_0 \ldots v_{m-2}$, for any $m \in \{1,2,\ldots,k-1\}$. We now establish that there exists a directed path from $v_{m-1}v_m \ldots v_{m-2}$ to a vertex of the form w. Notice that because $n \geq k+1$, we can choose a pair of elements, say, x and y, from $\overline{\{v_0, v_1, \ldots, v_{k-2}\}}$. The vertex $v_{m-1}v_m \ldots v_{k-2}v_0 \ldots v_{m-2}$ is adjacent to $v_m \ldots v_{k-2}v_0 \ldots \ldots v_{m-2}x$, which is, in turn, adjacent to $v_{m+1} \ldots v_{k-2}v_0 \ldots \ldots v_{m-2}xy$. Since there exists a directed path from a given vertex to each of its cyclic rotations, there exists a directed path from $v_{m+1} \ldots v_{k-2}v_0 \ldots v_{m-2}xy$ to $xyv_{m+1} \ldots v_{k-2}v_0 \ldots v_{m-2}$. The vertex $xyv_{m+1} \ldots v_{k-2}v_0 \ldots v_{m-2}$ is adjacent to $yv_{m+1} \ldots v_{k-2}v_0 \ldots v_{m-2}v_m$ which, in turn, is adjacent to $v_{m+1} \ldots v_{k-2}v_0 \ldots v_{m-2}v_mv_{m-1}$. Finally, this permutation is connected to its cyclic rotation $v_0v_1 \ldots v_{m-2}v_mv_{m-1}v_{m+1} \ldots v_{k-2}$. The pair $v_m < v_{m-1}$ now appears in the desired order in the permutation. We continue to perform this procedure until we reach a vertex of the form $w = w_0w_1 \ldots w_{k-2}$, where $w_0 < w_1 < \cdots < w_{k-2}$, to which v is connected.

The second part of the proof involves showing that $w = w_0w_1 \ldots w_{k-2}$, where $w_0 < w_1 < \cdots < w_{k-2}$, is connected to $12\ldots(k-1)$. We give an explicit proof, adding a case omitted by Jackson. We must consider two possibilities for the value of w_0. If $w_0 > 1$, then $w_i \neq i+1$, for all $0 \leq i \leq k-2$, due to the strictly increasing order of elements in w. Therefore, $w_0w_1 \ldots w_{k-2}$ is adjacent to $w_1 \ldots w_{k-2}1$, which is, in turn, adjacent to $w_2 \ldots w_{k-2}12$. By continuing to shift through w in this manner we will reach $12\ldots(k-1)$. If $w_0 = 1$, we must be careful in our approach. In this case, $w_0w_1 \ldots w_{k-2}$ is not adjacent to $w_1 \ldots w_{k-2}1$ because such an edge would yield an invalid "permutation" starting and ending with 1. Therefore, our first move is to a vertex $w_1 \ldots w_{k-2}x$, $x \in \overline{\{w_0, w_1, \ldots, w_{k-2}\}}$, adjacent to w. The vertex $w_1 \ldots w_{k-2}x$ is adjacent to $w_2 \ldots w_{k-2}x1$ since $w_0 < w_1 < \cdots < w_{k-2}$. The addition of x means that if there exists another $w_i = i+1$, for some $1 \leq i \leq k-2$, there is no problem since the value $i+1$ will be added to the end of a permutation one step after it is removed from the front of a permutation. We may now proceed, as described above, to shift through w until we reach $12\ldots(k-1)$.

Jackson does not show that $12\ldots(k-1)$ is connected to v. As this is not a trivial argument, we provide details here. To prove that $12\ldots(k-1)$ is connected to v, we use the same tools as above, but different reasoning. Denote $12\ldots(k-1)$ by $y_0y_1 \ldots y_{k-2}$. If $v_0 \notin \{y_0, y_1, \ldots, y_{k-2}\}$, then $y_0y_1 \ldots y_{k-2}$ is adjacent to $y_1y_2 \ldots y_{k-2}v_0$. Otherwise, $v_0 = y_i$, for some $0 \leq i \leq k-2$ and, by cyclic rotation, $y_0y_1 \ldots y_{k-2}$ is adjacent to $y_{i+1} \ldots y_{k-2}y_0 \ldots y_i = y_{i+1} \ldots y_{k-2}y_0 \ldots v_0$. Without loss of generality, assume we are in the second case. If $v_1 \notin \{y_0, y_1, \ldots, y_{k-2}\}$, then $y_{i+1} \ldots y_{k-2}y_0 \ldots v_0$ is adjacent to $y_{i+2} \ldots y_{k-2}y_0 \ldots v_0v_1$. Otherwise, $v_1 = y_j$, for some $0 \leq j \leq k-2$, $j \neq i$. Assume $j > i$. The vertex $y_{i+1} \ldots y_j \ldots y_{k-2}y_0 \ldots v_0$ is adjacent to $y_j \ldots y_{k-2}y_0 \ldots v_0 \ldots y_{j-1} = v_1 \ldots y_{k-2}y_0 \ldots v_0 \ldots y_{j-1}$. There exists $x \in \overline{\{y_0, y_1, \ldots, y_{k-2}\}}$ such that $v_1y_{j+1} \ldots \ldots y_{k-2}y_0 \ldots v_0 \ldots y_{j-1}$ is adjacent to $y_{j+1} \ldots y_{k-2}y_0 \ldots v_0 \ldots y_{j-1}x$. By cyclic rotation, this vertex is adjacent to $y_{i+1} \ldots y_{j-1}xy_{j+1} \ldots y_{k-2}y_0 \ldots v_0$, which is, in turn, adjacent to $y_{i+2} \ldots y_{j-1}xy_{j+1} \ldots v_0v_1$. We

Table 2.18 The representation of permutations of [4] given by the Ucycle 123415342154213541352435 with $M = 5$ (read column-wise)

Ucycle Representation $\Longleftrightarrow$ Permutation	
1234 $\Longleftrightarrow$ 1234	2135 $\Longleftrightarrow$ 2134
2341 $\Longleftrightarrow$ 2341	1354 $\Longleftrightarrow$ 1243
3415 $\Longleftrightarrow$ 2314	3541 $\Longleftrightarrow$ 2431
4153 $\Longleftrightarrow$ 3142	5413 $\Longleftrightarrow$ 4312
1534 $\Longleftrightarrow$ 1423	4135 $\Longleftrightarrow$ 3124
5342 $\Longleftrightarrow$ 4231	1352 $\Longleftrightarrow$ 1342
3421 $\Longleftrightarrow$ 3421	3524 $\Longleftrightarrow$ 2413
4215 $\Longleftrightarrow$ 3214	5243 $\Longleftrightarrow$ 4132
2154 $\Longleftrightarrow$ 2143	2435 $\Longleftrightarrow$ 1324
1542 $\Longleftrightarrow$ 1432	4351 $\Longleftrightarrow$ 3241
5421 $\Longleftrightarrow$ 4321	3512 $\Longleftrightarrow$ 3412
4213 $\Longleftrightarrow$ 4213	5123 $\Longleftrightarrow$ 4123

continue in this manner until we reach vertex v. Therefore, we conclude that $T_{n,k}$ is strongly connected.

To see that the in-degree of each vertex is equal to its out-degree, consider the vertex $v = v_0v_1 \dots v_{k-2} \in V(T_{n,k})$. This $(k-1)$-permutation can be extended to a k-permutation by adding one element that does not already appear in v. There are $n-(k-1) = n-k+1$ such elements. Let v' denote a vertex created by adding a new element to the beginning of v and then dropping the element v_{k-2}. There is an edge from each v' to v in $T_{n,k}$; therefore, the in-degree of v is $n-k+1$. Similarly, let v'' denote a vertex created by adding a new element to the end of v and then dropping the element v_0. There is an edge from v to each v'' in $T_{n,k}$; therefore, the out-degree of v is $n-k+1$. We have shown that both conditions of Theorem 2.5 are satisfied; therefore, $T_{n,k}$ is Eulerian. □

Ruskey and Williams have recently given an explicit construction for Ucycles of $(n-1)$-permutations of $[n]$ [43].

We now turn to Ucycles for n-permutations of $[n]$, which we will simply refer to as permutations and denote by S_n. There are at least four possible ways to define a Ucycle for permutations. The first was suggested by Chung et al. in [10]. A permutation of order n, $\pi = \pi_1 \dots \pi_n$, is **order isomorphic** to a sequence of length n, $s_1 \dots s_n$, if for $1 \leq i, j \leq n$, $\pi_i < \pi_j$ if and only if $s_i < s_j$. We will use the symbol $\sim$ to indicate that two permutations are order isomorphic. A Ucycle for permutations of $[n]$ is a circular sequence, $U = x_0 \dots x_{n!-1}$ of symbols from $\{1, \dots, M\}$ in which every permutation is order isomorphic to some contiguous subsequence of length n. The aim is to find the smallest M such that an order-isomorphic Ucycle exists. For $n > 2$, M must be at least $n+1$. It was conjectured by Chung et al. that there exist Ucycles for permutations of $[n]$ with $M = n+1$, and recently, this conjecture was proved by Johnson [35]. The sequence 123415342154213541352435 is a Ucycle for permutations of [4] with $M = 5$. The list of permutations represented by this sequence is shown in Table 2.18.

Theorem 2.24 ([35]). *For all $n \geq 3$ there exists a word W of length $n!$ over the alphabet $\{0,1,2,\ldots,n\}$ such that each element of S_n is order isomorphic to exactly one of the $n!$ cyclic intervals of length n in W.*

Johnson's proof uses a transition digraph, which we describe in the following paragraph. Note that it is a different transition digraph than defined in previous work on Ucycles for permutations [10] and k-permutations of $[n]$ [34]. In [10], the transition digraph for permutations is defined as follows: the vertices are the permutations of $[n]$, and there is an edge from one vertex to another if the $(n-1)$-tail of the first is order isomorphic to the $(n-1)$-head of the second. That is, there is an edge from $x_0x_1\ldots x_{n-1}$ to $y_0y_1\ldots y_{n-1}$ if $x_1\ldots x_{n-1}$ is order isomorphic to $y_0\ldots y_{n-2}$. As seen in the proof of Theorem 2.23, Jackson has another definition of a transition digraph for k-permutations of $[n]$, where $k \leq n+1$. See Fig. 2.15 for an illustration of this transition digraph.

Johnson defines his transition digraph for permutations as follows. Let $\{0,1,2,\ldots,n\}$ be the alphabet used for the Ucycle for S_n. The vertices are the n-permutations of $\{0,1,2,\ldots,n\}$, and there exists a directed edge from vertex $x = x_0\ldots x_{n-1}$ to vertex $y = y_0\ldots y_{n-1}$ if $x_{i+1} = y_i$, for all $0 \leq i < n-1$ (i.e. vertices are related in the same way as in other transition digraphs, but the alphabet is different). Thus, every vertex in the graph has both in-degree and out-degree equal to two.

To prove that order-isomorphic Ucycles for permutations which require $n+1$ elements exist, show there exists a directed cycle in the transition digraph of length $n!$ such that for each $a \in S_n$, there is some vertex of the cycle which is order isomorphic to a. Note that a Hamilton cycle is not required.

Johnson describes his proof approach as finding a collection of short cycles in the transition digraph which between them contain one vertex from each order-isomorphism class, then joining them up using an induction step. The method of joining cycles to create the final complete Ucycle is a similar idea to that used in proving the existence of Ucycles for symmetric BIBDs (see Chap. 5). We reproduce Johnson's proof here nearly in its entirety.

Proof (Proof of Theorem 2.24). We begin with some necessary machinery. Define a map on the integers:

$$s_x(i) = \begin{cases} i & \text{if } i < x, \\ i+1 & \text{if } i \geq x. \end{cases}$$

Slightly abusing this notation, write s_x for the map constructed by applying the above map coordinate-wise to an n-tuple, that is, $s_x(a_1a_2\ldots a_n) = s_x(a_1)s_x(a_2)\ldots s_x(a_n)$. This definition means that if $a = a_1a_2\ldots a_n \in S_n$ is a permutation of $[n]$ and $x \in [n+1]$, then $s_x(a)$ is the unique n-tuple of elements of $[n+1] \setminus \{x\}$ which is order isomorphic to a. This is the definition we need even though the final construction will produce a cycle for permutations of $[n]$ using the alphabet $\{0,1,\ldots,n\}$. Define a map r on the n-tuples which permutes the elements of the n-tuple cyclically. That is,

$$r(a_1a_2\ldots a_n) = a_2a_3\ldots a_{n-1}a_na_1.$$

Note that $(a, r(a))$ is an edge of the transition digraph and that $r^n(a) = a$.

Step 1: Finding short cycles in the transition digraph, G_n. The first step is to find a collection of short cycles (each of length n) in G_n which between them contain exactly one element from each order-isomorphism class of S_n. These cycles will use only n elements from the alphabet, and we will think of each cycle as being "labelled" with the remaining unused element. Suppose that for each $a = a_1a_2 \ldots a_{n-1} \in S_{n-1}$ we choose a label $\ell(a)$ from $[n]$. Let $0a$ be the n-tuple $0a_1a_2 \ldots a_{n-1}$. We have the following cycle in G_n:

$$s_{\ell(a)}(0a), r(s_{\ell(a)}(0a)), r^2(s_{\ell(a)}(0a)), \ldots, r^{n-1}(s_{\ell(a)}(0a)).$$

Denote this cycle by $C(a, \ell(a))$. As an example, $C(42135, 2)$ is the following cycle:

$$053146 \to 531460 \to 314605 \to 146053 \to 460531 \to 605314 \to 053146.$$

Note that for any choice of labels (i.e. any map ℓ), the cycles $C(a, \ell(a))$ and $C(b, \ell(b))$ are disjoint when $a, b \in S_n$ are distinct. Consequently, whatever the choice of labels, the collection of cycles

$$\cup_{a \in S_{n-1}} C(a, \ell(a))$$

is a disjoint union. It is easy to see that the vertices on these cycles contain between them exactly one n-tuple order isomorphic to each permutation in S_n.

We must now show how, given a suitable labelling, we can join up these short cycles.

Step 2: Joining two of these cycles. Suppose that $C_1 = C(a, x)$ and $C_2 = C(b, y)$ are two of the cycles in the transition digraph described above. What conditions on a, b and their labels x, y will allow us to join these cycles?

We may assume that $x \le y$. Suppose further that $1 \le x \le y - 2 \le n - 1$ and that a and b satisfy the following:

$$b_i = \begin{cases} a_i & \text{if } 1 \le a_i \le x - 1 \\ a_i + 1 & \text{if } x \le a_i \le y - 2 \\ x & \text{if } a_i = y - 1 \\ a_i & \text{if } y \le a_i \le n - 1. \end{cases}$$

If this happens, we will say that the pair of cycles $C(a,x)$, $C(b,y)$ are **linkable**.

In this case, $s_x(0a)$ and $s_y(0b)$ agree in all but one position; they differ only at the t for which $a_t = y - 1$ and $b_t = x$. It follows that there is a directed edge in the transition digraph from

$$r^t(s_x(0a)) = s_x(a_t \ldots a_{n-1}0a_1 \ldots a_{t-1})$$

to

$$r^{t+1}(s_y(0b)) = s_y(b_{t+1} \ldots b_{n-1}0b_1 \ldots b_t).$$

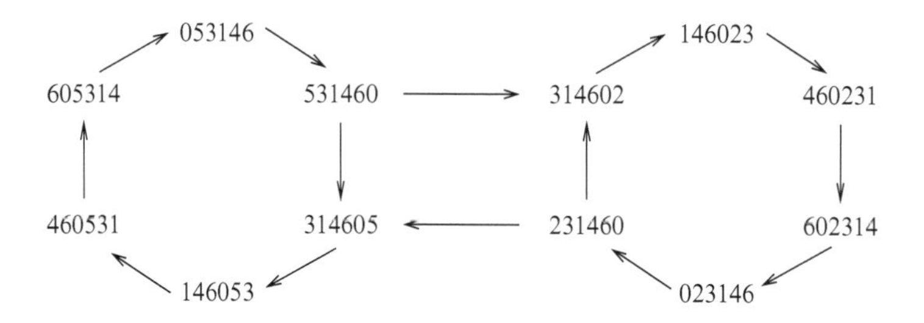

Fig. 2.16 The cycles $C(42135,2)$ and $C(23145,5)$ are linkable

Similarly, there is a directed edge in the transition digraph from

$$r^t(s_y(0b)) = s_x(b_t \dots b_{n-1}0b_1 \dots b_{t-1})$$

to

$$r^{t+1}(s_x(0a)) = s_y(a_{t+1} \dots a_{n-1}0a_1 \dots a_t).$$

If we add these edges to $C_1 \cup C_2$ and remove the edges

$$(r^t(s_x(0a)), r^{t+1}(s_x(0a)))$$

and

$$(r^t(s_y(0b)), r^{t+1}(s_y(0b))),$$

then we produce a single cycle of length $2n$ whose vertices are precisely the vertices in $C_1 \cup C_2$.

Note that if $x = y - 1$, then the other conditions imply that $a = b$, and so although we can perform a similar linking operation, it is not useful. If $x = y$, then b is not well defined.

As an example of the linking operation, consider the linkable pair of 6-cycles $C(42135,2)$ and $C(23145,5)$ in G_6. If we add the edges $531460 \to 314602$ and $231460 \to 314605$ and remove the edges $531460 \to 314605$ and $231460 \to 314602$, then a single cycle of length 12 in G_6 is produced. See Fig. 2.16.

Step 3: Joining all these cycles. We now show that this linking operation can be used repeatedly to join a collection of disjoint short cycles—one for each $a \in S_{n-1}$—together.

Let $H_n = (V,E)$ be the (undirected) graph with,

$$V = \{(a,x) : a \in S_{n-1}, x \in [n]\}$$

$$E = \{((a,x),(b,y)) : C(a,x), C(b,y) \text{ are linkable}\}$$

If we can find a subtree T_n of H_n (having $(n-1)!$ vertices) which contains exactly one vertex (a,x) for each $a \in S_{n-1}$, then we will be able to construct the required cycle. Take any vertex $(a,\ell(a))$ of T_n and consider the cycle

$C(a,\ell(a))$ associated with it. Consider also the cycles associated with all the neighbours of $(a,\ell(a))$ in T_n. The linking operation described above can be used to join the cycles associated with these neighbours in $C(a,\ell(a))$. This is because the definition of adjacency in H_n guarantees that we can join each of these cycles individually. Also, the fact that every vertex in G_n has out-degree two means that the joining happens at different places along the cycle. That is, if $(b,\ell(b))$ and $(c,\ell(c))$ are distinct neighbours of $(a,\ell(a))$, then the edge of $C(a,\ell(a))$ which must be deleted to join $C(b,\ell(b))$ to it is not the same as the one which must be deleted to join $C(c,\ell(c))$ to it. We conclude that we can join all of the relevant cycles to the cycle associated with $(a,\ell(a))$. The connectivity of T_n now implies that we can join all of the cycles associated with vertices of T_n into one cycle. This is plainly a cycle with the required properties.

The next step is to find such a subtree in H_n.

Step 4: Constructing a suitable tree. Johnson proves, by induction on n, the stronger statement that, for all $n \geq 5$, there is a subtree T_n of H_n of order $(n-1)!$ which satisfies the following seven properties. For a tree satisfying property 1, we denote the unique vertex in $V(T_n)$ of the form (a,x) by $v(a)$:

1. For all $a \in S_{n-1}$, there exists a unique $x \in [n]$ such that $(a,x) \in V(T_n)$.
2. $(12\ldots(n-1),1) \in V(T_n)$.
3. $(23\ldots(k-1)1(k)(k+1)\ldots(n-1),k) \in V(T_n)$ for all $3 \leq k \leq n$.
4. $(32145\ldots(n-1),2) \in V(T_n)$.
5. $(243156\ldots(n-1),3) \in V(T_n)$.
6. $v(31245\ldots(n-1))$ is a leaf in T_n.
7. $v(24135\ldots(n-1))$ is a leaf in T_n.

For $n = 5$ a suitable tree can be found. Suppose that $n \geq 5$ and that we have a subtree T_n of the graph H_n which satisfies the above conditions; use this to build a suitable subtree of H_{n+1}. As this part of the construction is somewhat involved and does not shed light on the construction of Ucycles for block designs, we direct the interested reader to the details in [35]. □

We now turn to a second idea of equivalence for permutations. Let 1_m denote the vector of m ones. Let $\Pi^k_{i,j}$ denote the set of all k-permutations of $\{i,i+1,\ldots,j\}$ and write elements of $\Pi^k_{i,j}$ as length k vectors of distinct terms from $\{i,i+1,\ldots,j\}$. Define the following equivalence relation on $\Pi^m_{0,n}$. For $\mathbf{a},\mathbf{b} \in \Pi^m_{0,n}$, $\mathbf{a} \sim \mathbf{b}$ if and only if $\mathbf{a}-\mathbf{b} \equiv k \cdot 1_m \pmod{n+1}$ for some $k \in \mathbb{N}$. There are $n!/(n-m)!$ equivalence classes of the elements of $\Pi^m_{0,n}$, each of which corresponds to an element of $\Pi^m_{1,n}$. This means we have a representation of permutations of $[n]$ that can be viewed as equivalence classes of n-permutations of $[n+1]$. Define an **equivalence class Ucycle** to be a sequence of length $n!$ such that amongst the length n subsequences, each equivalence class is represented exactly once. Hurlbert and Isaak have shown that, given this equivalence relation, there exist equivalence class

Ucycles representing $\Pi_{1,n}^{n}$ [32]. These equivalence class Ucycles use $n+1$ symbols to represent Ucycles for permutations of $[n]$.

Theorem 2.25 ([32]). *There exists a complete family of equivalence class Ucycles for permutations of* $\{1,2,\ldots,n\}$ *using the symbols* $\{0,1,\ldots,n\}$.

A third way to define a Ucycle for permutations requires that the original alphabet be maintained; however, it relaxes the condition that each permutation of order n appears exactly once. Using this definition, the goal is to find the shortest circular sequence of symbols from $[n]$ such that every permutation of $[n]$ appears as a contiguous subsequence *at least* once.

Finally, a fourth method for representing permutations in a Ucycle is by using the original alphabet of the objects, but using representatives of rank $n-1$, where the missing symbol is implied. These have been called **shorthand Ucycles** by Williams [58], though the concept of using a minimum rank to represent combinatorial objects was introduced previously. Jackson proved that such Ucycles exist for all n [34], and recently Williams and Ruskey have given an explicit construction for Ucycles for $(n-1)$-permutations of an n-set using an iterative, loopless algorithm that requires only $O(n)$ space [43]; this work was extended in [27]. They note that every $(n-1)$-permutation can be uniquely extended to a permutation of the n-set by appending the unique missing symbol. Further, Williams and his colleagues have proven the existence of shorthand Ucycles for permutations of multisets [58] and words with restricted weights [42,46,47].

2.3.4.2 *k*-subsets of *n*-sets

The largest body of work on Ucycles is focused on determining the existence of Ucycles for k-subsets of n-sets. This is due, in part, to the fact that it has thus far proved to be a much more difficult question to answer than the existence of Ucycles for permutations and partitions. Throughout this section, we assume $1<k<n$, as the two extreme cases are trivial. As a result, the alphabet used in Ucycles for k-subsets of $[n]$ is exactly the n-set. It seems this would make the problem easier to handle; however, we will see that the existence of Ucycles for k-subsets of n-sets is known only for some specific values of k.

Because we are dealing with subsets, once a given subset has been represented in the Ucycle, it may not appear in any other order elsewhere in the sequence. This means that determining a transition digraph is difficult as the vertices and edges are dependent on the part of the Ucycle that has already been fixed. For example, if the subset $\{1,2,3\}$ is represented in the Ucycle as 123, then the arc $\{1,2,3\} \rightarrow \{2,3,4\}$ exists in the transition digraph. However, if the subset is represented as 213, $\{1,2,3\} \rightarrow \{2,3,4\}$ cannot be an arc in the transition digraph. The following modular condition is easily established.

Lemma 2.3 ([10]). *If there is a Ucycle for k-subsets of* $[n]$, *then k divides* $\binom{n-1}{k-1}$.

Ucycles exist for $k=2$ as long as the condition of Lemma 2.3 is satisfied, that is, as long as n is odd. These are simply Euler tours in the complete graph K_n.

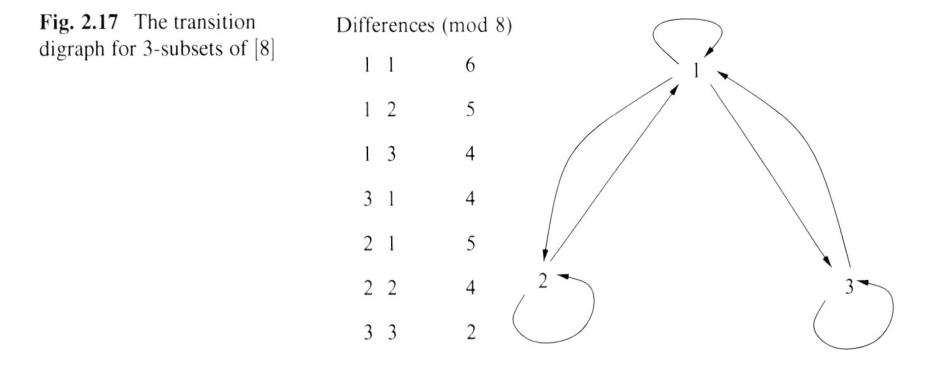

Fig. 2.17 The transition digraph for 3-subsets of $[8]$

Ucycles for the ordered distinct pairs of an n-set similarly correspond to Euler tours of the complete directed graph. Loops can even be added to include *all* ordered pairs. Ucycles for ordered pairs with additional requirements have application in statistical design. These objects are discussed in Sect. 6.5.2.

The following two theorems are due to Jackson.

Theorem 2.26 ([34]). *There exist Ucycles for* 3*-subsets of* $[n]$*, when* $n \geq 8$*, if and only if* $\binom{n-1}{2} \equiv 0 \pmod{3}$.

Before presenting a proof of Theorem 2.26, we introduce some helpful machinery. Let $X = \{x_1, x_2, \ldots, x_k\}$ be a k-subset of $[n]$ written such that $x_1 < x_2 < \cdots < x_k$. To visualize the following idea, think of the integers from 0 to $n-1$ arranged around a circle. Two k-subsets are of the same **type** if they differ by just a rotation around the circle. For example, 123 is of the same type as 345. Formally, we can define the **type** of X to be $(x_2 - x_1, x_3 - x_2, \ldots, x_k - x_{k-1}, x_1 - x_k)$, with arithmetic performed modulo n. An ordered k-subset can be uniquely identified by its first entry and (the first $k-1$ entries of) its type. The set of all k-subsets of $[n]$ of a certain type is denoted by omitting the largest difference that is unequal to any of the other $k-1$ differences.

Proof (Proof of Theorem 2.26). To prove that Ucycles for 3-subsets of $[n]$ exist, a similar argument to that involved in proving the existence of Ucycles for permutations is employed. A subsequence of k elements in a Ucycle with differences $d_1 d_2 \ldots d_{k-1}$ (we have dropped the brackets and commas between differences and we ignore the difference between the last and the first elements) must be followed by a subsequence of k elements with differences $d_2 \ldots d_{k-1} d_k$. A transition digraph can be created for these difference types with vertices representing vectors of length $k-2$ and each edge representing a length $k-1$ difference type (which in turn represents many possible k-subsets). Given vertices $d_1 \ldots d_{k-2}$ and $d_2 \ldots d_{k-1}$, there is a directed edge between them representing the difference type $d_1 \ldots d_{k-1}$.

For 3-subsets, the difference types are of the form (1) three unequal differences, or (2) two equal differences and a third number different from these two (but never three equal differences as $\gcd(n,3) = 1$). Since we can drop one of the differences without any loss of information, the types are written as (1) ij or ji, where $i \neq j$, or (2) ii. The transition digraph for 3-subsets of $[8]$ is pictured in Fig. 2.17. The

Fig. 2.18 A Ucycle for 3-subsets of [8]

```
1 2 3 5 7 8 3 6 7 8 2 4 5 8 3 4 5 7 1 2 5 8
6                                         1
3                                         2
2                                         4
8                                         6
6                                         7
5                                         2
4 1 6 5 3 1 8 7 4 1 8 6 4 3 2 7 4 3 1 7 6 5
```

possible difference types modulo 8 are also listed, with a space separating the two differences used to represent each type (we always drop the largest difference that appears exactly once).

For $n = 8$, the approach to finding a Ucycle for 3-subsets of $[8]$ reduces to the following. An Euler tour in the transition digraph defines a circular arrangement that contains each difference type uniquely (as a 2-set). The sum of the differences in the circular arrangement is coprime to n; therefore, a list which contains each 3-subset of $[8]$ exactly once can be obtained by starting with a given initial value in $[8]$ and successively adding this list of differences n times before returning to the starting point. Figure 2.18 is a Ucycle for 3-subsets of an 8-set obtained by taking the difference list 11221331 (an Euler tour in Fig. 2.17) and successively adding this list of differences with starting point 1.

When $n > 8$ a little more work is required. Recall that 3-subsets can be represented by their first element and two differences; thus, $(x, d_1 d_2)$ represents the subset $\{x, x + d_1, x + d_1 + d_2\}$. In a Ucycle, the 3-subset represented by $(x, d_1 d_2)$ is followed by the 3-subset represented by $(x + d_1, d_2 d_3)$. Define an expanded transition digraph with vertices (x, d) such that the edge from vertex (x, d_1) to $(x + d_1, d_2)$ is the 3-subset representative $(x, d_1 d_2)$. Each vertex has the same in-degree and out-degree, and one can show that the graph is also strongly connected; therefore, appealing to Theorem 2.5, there is an Euler tour in the expanded transition digraph. □

Using a similar approach to that used in proving Theorem 2.26, Jackson has also partially determined the existence of Ucycles for 4-subsets.

Theorem 2.27 ([34]). *There exist Ucycles for 4-subsets of* $[n]$, $n \geq 9$, *provided* $\binom{n-1}{3} \equiv 0 \pmod{4}$ *and* $\gcd(n, 4) = 1$.

Note that there are values of n, namely, $n \equiv 2 \pmod{8}$, that satisfy the modular condition of Lemma 2.3, but are not covered by Theorem 2.27. These cases where n is not coprime to four remain unresolved.

Using much the same approach as Jackson, Hurlbert has also shown that Ucycles exist for 3-subsets of $[n]$ and 4-subsets of $[n]$, when n and k are coprime [30]. In addition, he has proved the existence of Ucycles for 6-subsets of n when $\gcd(n, 6) = 1$ and n is sufficiently large. Nothing is known about the existence of Ucycles for k-subsets of $[n]$ for $k = 5$ or for $k \geq 7$, although computer searches have constructed some specific examples of Ucycles for higher k. Chung et al. have made the following conjecture, suggesting that the statement of Lemma 2.3 is both necessary and sufficient.

Conjecture 2.1 ([10]). There exist Ucycles for k-subsets of $[n]$, provided k divides $\binom{n-1}{k-1}$ and n is sufficiently large.

However, Stevens et al. have shown that Ucycles do not exist for $(n-2)$-subsets of $[n]$ when $n \geq 4$ [50]. That is, n must be at least $k+3$ for a Ucycle representing the k-subsets of $[n]$ to exist. The approach in [50] is to determine structural features that these Ucycles would have if they existed, and then show that these features cannot possibly be realized.

As with permutations, the non-existence of a Ucycle for k-subsets of $[n]$ does not end the investigation. It is possible to search for how close one can get to such a Ucycle in two ways. One way is to find the longest sequence having all properties of a Ucycle, except not every subset will appear; this is called a **cyclic packing word**. The definition can be further relaxed to a non-cyclic sequence having Ucycle properties containing as many k-subsets of $[n]$ as possible. Such a sequence is called a **linear packing word**. The other way to get "close" to a Ucycle is to find the shortest sequence such that every subset appears at least once. Such a sequence is called a **cyclic covering word**. These are similar notions to packings and coverings in design theory.

Theorem 2.28 ([50]). *The longest possible cyclic packing word of $(n-2)$-subsets of $[n]$ has length n and a word achieving this bound of $[n]$ always exists:* $12\cdots(n-1)n$.

Theorem 2.29 ([50]). *The longest possible linear packing word of $(n-2)$-subsets of $[n]$ has length $3n-6$ and a word achieving this bound always exists:*

$$123\cdots(n-4)(n-3)(n-2)(n-1)123\cdots$$
$$\cdots(n-4)(n-3)n(n-1)123\cdots(n-5)(n-4).$$

Hurlbert et al. have shown that cyclic packing words that contain almost all the k-subsets of $[n]$ always exist.

Theorem 2.30 ([16]). *For any positive k, there exists a cyclic packing word of the k-subsets of an n-set of length*

$$(1-o(1))\binom{n}{k}.$$

Hurlbert extended the definition of Ucycles for k-subsets of $[n]$ to consider sequences in which every k-subset appears as a consecutive subsequence at least once and all k-subsets appear equally often. These are called ***t*-cover Ucycles**, where t is the number of times each subset appears. Naturally, these will be cyclic sequences of length $t \cdot \binom{n}{k}$. The following lemma is an extension of Lemma 2.3 to multicover Ucycles.

Lemma 2.4 ([31]). *If there is a t-cover Ucycle for k-subsets of* $[n]$*, then k divides* $t \cdot \binom{n-1}{k-1}$.

Hurlbert's goal is to find the smallest t, given n and k, such that a t-cover Ucycle exists. Denote this value by $U(n,k)$. Note that the known results for the existence of Ucycles for k-subsets of n-sets imply that for $k = 2,3,4,6$, n coprime to k and n sufficiently large, $U(n,k) = 1$.

Theorem 2.31 ([31]). $U(n,k) \leq k$ *for all* $n \geq k$.

In the case when k is a prime power and k divides n, the above inequality becomes an equality. To prove this corollary we require the following lemma due to E. Kummer stated in [29]. Given a prime p and an integer x, let $v_p(x)$ denote the maximum exponent r such that p^r divides x.

Lemma 2.5 (See [29]). $v_p(\binom{n}{k})$ *is equal to the number of borrows needed when subtracting k from n in base p.*

Corollary 2.2 ([17]). *If k is a prime power and k divides n, then* $U(n,k) = k$.

Proof. Suppose $k = p^q$, where p is prime, and suppose $n = mk$, $m \in \mathbb{N}$. We wish to determine the divisibility of $\binom{n-1}{k-1}$ by p. Written in base p

$$k = 1\underbrace{00\ldots0}_{q} \text{ and } n = x_1 \ldots x_i \underbrace{0\ldots0}_{q}.$$

So,

$$k-1 = \underbrace{(p-1)(p-1)\ldots(p-1)}_{q} \text{ and } n-1 = y_1 \ldots y_i \underbrace{(p-1)\ldots(p-1)}_{q}.$$

Therefore,

$$(n-1)-(k-1) = y_1 \ldots y_i \underbrace{0\ldots0}_{q},$$

and no borrows are required. Lemma 2.5 implies that zero is the largest exponent of p that divides $\binom{n-1}{k-1}$. Lemma 2.4 requires that $k = p^q$ divide $U(n,k) \cdot \binom{n-1}{k-1}$, so k must divide $U(n,k)$. Therefore, since $U(n,k) \leq k$, $U(n,k) = k$. □

Finally, we note some recent results of Hurlbert, Johnson and Zahl, regarding the existence of Ucycles for t-multisets of $[n]$.

Theorem 2.32 ([33]). *Let* $n_0(3) = 4, n_0(4) = 5$ *and* $n_0(6) = 11$*. Then, for* $t \in \{3,4,6\}$ *and* $n \geq n_0(t)$*, Ucycles for t-multisets of* $[n]$ *exist whenever n is relatively prime to t.*

Theorem 2.32 is proved using a transition digraph. Of particular interest is the fact that the authors are also able to prove the following result using an inductive method.

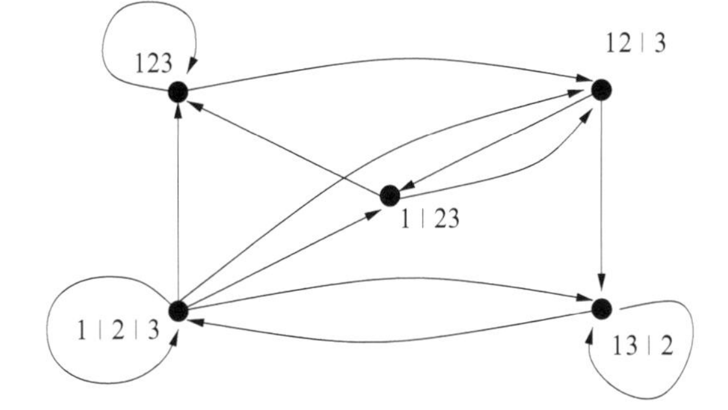

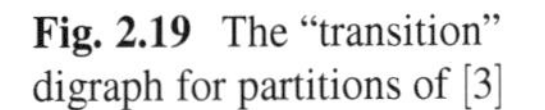
Fig. 2.19 The "transition" digraph for partitions of [3]

Theorem 2.33 ([33]). *For $t = 3$, Ucycles for the 3-multisets of $[n]$ exist provided $3 \nmid n$.*

2.3.4.3 Partitions of an n-set

Partitions of an n-set also require some adjustment to the definition of a Ucycle. The number of partitions of $[n]$ is the sum of the Stirling numbers of the second kind for n and all k between 1 and n. To represent a partition of $[n]$ we use a string of length n with elements from $A = \{a, b, c, \ldots\}$. We put $i, j \in [n]$ in the same class of the partition if and only if the ith and jth entries of the string are the same. For example, *abacbcdc* represents the partition $13|25|468|7$. Thus, *abcbccccddcdeec* is a Ucycle for the partitions of $[4]$. Create a "transition" digraph for the partitions of $[n]$ with vertices representing the partitions, and place a directed edge from vertex u to vertex v if the alphabetic sequence representing partition u can be followed (in a Ucycle) by the alphabetic sequence representing partition v. Figure 2.19 illustrates the "transition" digraph for partitions of $[3]$.

In order to find a Ucycle for the partitions of $[n]$ we must find a Hamilton cycle in this "transition" digraph. As for k-subsets of $[n]$, finding such a Hamilton cycle does not guarantee the existence of a Ucycle as it is necessary to "lift" the Hamilton cycle to a Ucycle by assigning appropriate symbols in order to realize the corresponding partitions. There is a Hamilton cycle in the digraph shown in Fig. 2.19, but it is not possible to lift it to a Ucycle. Suppose the Ucycle for partitions of $[3]$ is $x_0x_1x_2x_3x_4$. Given the Hamilton cycle 123, $12|3$, $13|2$, $1|2|3$, $1|23$, we find that $x_0 = x_1 = x_2$, $x_3 \neq x_2$, $x_4 = x_2$ and $x_0 \neq x_4$, which is a contradiction. In fact, there is no Ucycle for the partitions of $\{1,2,3\}$ [10]. However, Chung et al. claim (but do not explicitly prove) that it is possible to lift Hamilton cycles for $n \geq 4$ [10]. The problem with Hamilton cycles for $n = 3$ (and a potential problem in general) is that the inequalities implied by the lifting of the cycle result in $x \neq x$, for some x in the Ucycle. However, it is possible to prevent such an occurrence when $n \geq 4$ by requiring that a certain sequence of partitions occur in the cycle and that this sequence not contain the partition of $[n]$ into n parts.

Theorem 2.34 ([10,29]). *There exist Ucycles for partitions of* $[n]$ *when* $n \geq 4$.

Proof. We show the validity of this statement for partitions of $n \geq 8$ by exhibiting the necessary subsequence of partitions that ensures the avoidance of the $x \neq x$ condition:

(A) For n even, the sequence is $ab \ldots ba \ldots abac \ldots cded \ldots de \ldots ede$, and
(B) for n odd, the sequence is $ab \ldots bab \ldots babc \ldots cded \ldots ded \ldots ded$,

where $y \ldots y$ indicates a contiguous sequence of the letter y. The large blocks of the letters a, b, d and e have length $\lfloor (n-4)/2 \rfloor$, and the centre block of the letter c has length n. Note that this subsequence of the letter c represents the partition of $[n]$ into one part. Suppose the Ucycle looks like $x_0 x_1 \ldots x_{m-1}$ with x_i to the left of the centre block and x_j to the right of the centre block. Because the centre block is of length n there is no relation implied between x_i and x_j. The removal of the edges represented by sequence A or sequence B from the "transition" digraph does not disconnect it. It is possible to find a Hamilton cycle in the rest of the digraph and adjoin it to either sequence A or B in order to produce a Ucycle. □

One can also consider k-partitions of $[n]$, that is, partitions of $[n]$ into k classes, where $2 \leq k \leq n-1$. Little is known about these cases. Hurlbert has shown that Ucycles for k-partitions of $[n]$ exist for $n \geq 3$ odd and $k = n-1$ [29]. Casteels has shown that there is an algebraic way to enumerate all Ucycles of this form [8]. Furthermore, for n even, he has shown there do not exist Ucycles for $(n-1)$-partitions of $[n]$.

2.3.4.4 Other Combinatorial Objects

Williams has looked at Ucycles for several other types of combinatorial object. For example, Williams proved the following result for any finite language which includes all words with a given multiset of letters; a **fixed-content language**.

Theorem 2.35 ([58]). *Given a fixed-content language L, L has a shorthand Ucycle if and only if it has a cyclic Gray code where successive strings are obtained by right shifting the first symbol into the last or second last position.*

This work is applied to binary words with constrained weights in [42,46,47].

Another combinatorial object for which Ucycle results have been established is subsets of the binary strings of length n. The terminology used is that of de Bruijn sequences, in deference to the binary nature of the objects; however, we reserve the term de Bruijn for orderings of *all* 2^n n-strings in this monograph. Sawada, Stevens and Williams define a **maximum-density de Bruijn sequence** to be a circular binary sequence of length $\binom{n}{0} + \binom{n}{1} + \binom{n}{2} + \cdots + \binom{n}{m}$ that contains each binary string of length n with density (i.e. weight) between 0 and m, inclusive [46]. They efficiently generate maximum-density de Bruijn sequences for all values of n and m, which amounts to the following theorem:

Theorem 2.36 ([46]). *There exists a Ucycle containing all length n binary strings of weight between zero and m, for all $n \in \mathbb{N}$ and all $0 \leq m \leq n$.*

The algorithm for constructing the Ucycles of Theorem 2.36 is improved in [27]. In a similar vein, though using the idea of shorthand Ucycles, Ruskey, Sawada and Williams have constructed Ucycles for the n-bit strings of density d [47].

Theorem 2.37 ([47]). *There exists a Ucycle containing all length n binary strings of weight d, for all $n \in \mathbb{N}$ and $d \leq n$.*

They call these **fixed-density de Bruijn sequences**.

Recently, Brockman, Kay and Snively [4] have generalized the definition of Ucycle (by changing the representation from a list to a labelled graph to which "windows" are applied for reading off subgraphs) and obtained the following results:

Theorem 2.38 ([4]). *For each $k \geq 0$, $k \neq 2$, there exists a Ucycle for simple labelled graphs on k vertices.*

Theorem 2.39 ([4]). *For each $k \neq 2$, Ucycles exist for the following classes of graphs on k vertices: graphs with loops, graphs with multiple edges (with up to m duplications of each edge), directed graphs, hypergraphs and j-uniform hypergraphs.*

Theorem 2.40 ([4]). *Ucycles exist for trees on k vertices for $k \geq 3$.*

Conjectures

Conjecture 2.2 ([10]). There exist Ucycles for k-subsets of $[n]$, provided k divides $\binom{n-1}{k-1}$ and n is sufficiently large.

Conjecture 2.3 ([50]). For $n \geq 6$ even, it is not possible to have a length $n^2/2$ cyclic covering word for the $(n-2)$-subsets of an n-set.

Exercises and Problems

Exercise 2.1. Show how to recursively construct a $\mathrm{TS}(3v,\lambda)$ and $\mathrm{TS}(2v+1,\lambda)$ from a $\mathrm{TS}(v,\lambda)$.

Exercise 2.2. Determine the block-intersection graph of the $\mathrm{STS}(9)$. Is this graph Hamiltonian?

Exercise 2.3. Show that equation (2.1) defines the standard binary reflected Gray code.

Exercise 2.4. Prove Theorem 2.20.

Exercise 2.5. Prove that Martin's construction, given on page 42, does indeed construct a de Bruijn sequence of order n.

Exercise 2.6. Suppose that a Ucycle for the $(n-2)$-subsets of an n-set exists.

1. Show that it must have a subword of length n that is a permutation of the n-set.
2. Show that any given $(n-4)$-subset can appear at most twice and determine, up to isomorphism, the two preceding and two postceding symbols for each $(n-4)$-subset.

Exercise 2.7. Show that if a packing word for the $(n-2)$-subsets of an n-set exists, then the positions of two occurrences of the same symbol, with no other occurrence of the symbol between them, must differ by at least $n-1$ and at most $2n-3$.

Exercise 2.8. Show that if a packing word for the $(n-2)$-subsets of an n-set exists, a subword of length n which is a permutation of the n-set appears at position i, and the packing word is of length at least $2n+i-3$, then the symbols that appear in positions $2n-2+j$ and $n-1+j$ are identical for all $0 \leq j \leq i-1$.

Exercise 2.9. Prove Theorems 2.28 and 2.29.

Problem 2.1. Investigate the structural necessities for the existence of a Ucycle of $(n-3)$-subsets of an n-set.

References

1. Beezer, R.A.: Counting configurations in designs. J. Combin. Theory Ser. A **96**(2), 341–357 (2001)
2. Bhat, G.S., Savage, C.D.: Balanced Gray codes. Electron. J. Combin. **3**(1), Research Paper 25, (electronic, 11 pp.) (1996)
3. Bitner, J.R., Ehrlich, G., Reingold, E.M.: Efficient generation of the binary reflected Gray code and its applications. Comm. ACM **19**(9), 517–521 (1976)
4. Brockman, G., Kay, B., Snively, E.E.: On universal cycles of labeled graphs. Electron. J. Combin. **17**(1), Research Paper 4, (electronic, 9 pp.) (2010)
5. Brualdi, R.A.: Introductory Combinatorics. Prentice-Hall, Upper Saddle River, NJ (1999)
6. Bryant, D., El-Zanati, S.: Graph Decompositions, chap. VI.24, pp. 477–486. In: Colbourn and Dinitz [14] (2007)
7. Buck, M., Wiedemann, D.: Gray codes with restricted density. Discrete Math. **48**, 163–171 (1984)
8. Casteels, K.: Universal cycles for $(n-1)$-partitions of an n-set. Master's thesis, Carleton University, Ottawa, ON (2004)
9. Chase, P.J.: Combination generation and Graylex ordering. Congr. Numer. **69**, 215–242 (1989)
10. Chung, F., Diaconis, P., Graham, R.: Universal cycles for combinatorial structures. Discrete Math. **110**, 43–59 (1992)
11. Chvátal, V., Erdős, P.: A note on Hamiltonian circuits. Discrete Math. **2**, 111–113 (1972)
12. Colbourn, C.J.: Combinatorial aspects of covering arrays. Matematiche (Catania) **59**(1–2), 125–172 (2004)
13. Colbourn, C.J.: Covering Arrays, chap. VI.10, pp. 361–365. In: Colbourn and Dinitz [14] (2007)

14. Colbourn, C.J., Dinitz, J.H. (eds.): Handbook of Combinatorial Designs, second edn. Chapman & Hall/CRC, Boca Raton, FL (2007)
15. Colbourn, C.J., Rosa, A.: Triple Systems. Oxford Mathematical Monographs. The Clarendon Press Oxford University Press, New York (1999)
16. Curtis, D., Hines, T., Hurlbert, G., Moyer, T.: Near-universal cycles for subsets exist. SIAM J. Discrete Math. **23**(3), 1441–1449 (2009)
17. Dewar, M.: Gray codes, universal cycles and configuration orderings for block designs. Ph.D. thesis, Carleton University, Ottawa, ON (2007)
18. Dewar, M., Proos, J., McInnes, L.: Monotone Gray codes for vectors of the form $[-m,m]^k$ and $[0,m]^k$. Submitted to Electron. J. Combin. (2012)
19. Eades, P., Hickey, M., Read, R.C.: Some Hamilton paths and a minimal change algorithm. J. Assoc. Comput. Mach. **31**(1), 19–29 (1984)
20. Eades, P., McKay, B.: An algorithm for generating subsets of fixed size with a strong minimal change property. Inform. Process. Lett. **19**(3), 131–133 (1984)
21. Fredricksen, H.: A survey of full length nonlinear shift register cycle algorithms. SIAM Rev. **24**(2), 195–221 (1982)
22. Gordon, D.M., Stinson, D.R.: Coverings, chap. VI.11, pp. 365–373. In: Colbourn and Dinitz [14] (2007)
23. Grannell, M.J., Griggs, T.S.: Configurations in Steiner triple systems. In: Combinatorial Designs and their Applications (Milton Keynes, 1997), vol. 403, pp. 103–126. Chapman & Hall/CRC, Boca Raton, FL (1999)
24. Grannell, M.J., Griggs, T.S., Mendelsohn, E.: A small basis for four-line configurations in Steiner triple systems. J. Combin. Des. **3**(1), 51–59 (1995)
25. Grannell, M.J., Griggs, T.S., Whitehead, C.A.: The resolution of the anti-Pasch conjecture. J. Combin. Des. **8**(4), 300–309 (2000)
26. Hartman, A.: Software and hardware testing using combinatorial covering suites. In: Graph Theory, Combinatorics and Algorithms, pp. 237–266. Springer, New York (2005)
27. Holroyd, A., Ruskey, F., Williams, A.: Shorthand universal cycles for permutations. Algorithmica, **64**, 215–245 (2012)
28. Horák, P., Rosa, A.: Decomposing Steiner triple systems into small configurations. Ars Combin. **26**, 91–105 (1988)
29. Hurlbert, G.H.: Universal cycles: on beyond de Bruijn. Ph.D. thesis, Rutgers University, New Brunswick, NJ (1990)
30. Hurlbert, G.H.: On universal cycles for k-subsets of an n-set. SIAM J. Disc. Math. **7**(4), 598–604 (1994)
31. Hurlbert, G.H.: Multicover Ucycles. Discrete Math. **137**, 241–249 (1995)
32. Hurlbert, G.H., Isaak, G.: Equivalence class universal cycles for permutations. Discrete Math. **149**, 123–129 (1996)
33. Hurlbert, G.H., Johnson, T., Zahl, J.: On universal cycles for multisets. Discrete Math. **309**, 5321–5327 (2009)
34. Jackson, B.W.: Universal cycles of k-subsets and k-permutations. Discrete Math. **117**, 141–150 (1993)
35. Johnson, J.R.: Universal cycles for permutations. Discrete Math. **309**, 5264–5270 (2009)
36. Knuth, D.E.: The Art of Computer Programming, vol. 4. Addison-Wesley, Upper Saddle River, NJ (2005)
37. Kompel′maher, V.L., Liskovec, V.A.: Successive generation of permutations by means of a transposition basis. Kibernetika (Kiev) (3), 17–21 (1975)
38. Kreher, D.L., Stinson, D.R.: Combinatorial Algorithms: generation, Enumeration and Search. CRC Press, Boca Raton, FL (1999)
39. Martin, M.H.: A problem in arrangements. Bull. Amer. Math. Soc. **40**, 859–864 (1934)
40. Ruskey, F.: Adjacent interchange generation of combinations. J. Algorithms **9**(2), 162–180 (1988)

41. Ruskey, F.: Simple combinatorial Gray codes constructed by reversing sublists. In: Proceedings of the Fourth International Symposium (ISAAC '93) held in Hong Kong, Dec 15–17, 1993. Edited by K.W. Ng, P. Raghavan, N.V. Balasubramanian and F.Y.L. Chin. Lecture Notes in Computer Science, vol. 762, Springer-Verlag, Berlin, pp. xiv+542 (1993). ISBN: 3-540-57568-5
42. Ruskey, F., Sawada, J., Williams, A.: Fixed-density de Bruijn sequences. SIAM J. Disc. Math. **26**(2), 605–617 (2012)
43. Ruskey, F., Williams, A.: An explicit universal cycle for the $(n-1)$-permutations of an n-set. ACM Trans. Algorithms **6**(3), Art. 45, (electronic, 12 pp.) (2010)
44. Savage, C.D.: A survey of combinatorial Gray codes. SIAM Rev. **39**(4), 605–629 (1997)
45. Savage, C.D., Winkler, P.: Monotone Gray codes and the middle levels problem. J. Combin. Theory. Ser. A **70**(2), 230–248 (1995)
46. Sawada, J., Stevens, B., Williams, A.: De Bruijn sequences for the binary strings with maximum density. In: Proceedings of the 5th international conference on WALCOM: Algorithms and computation, pp. 182–190. Springer-Verlag, Berlin, Heidelberg (2011)
47. Sawada, J., Williams, A.: A Gray code for fixed-density necklaces and Lyndon words in constant amortized time. To appear in Theoret. Comput. Sci. (2011)
48. Slater, P.J.: Generating all permutations by graphical transpositions. Ars Combin. **5**, 219–225 (1978)
49. Sloane, N.J.A.: The On-Line Encyclopedia of Integer Sequences. http://www.research.att.com/~njas/sequences/ (2004). Accessed June 2004
50. Stevens, B., Buskell, P., Ecimovic, P., Ivanescu, C., Malik, A., Savu, A., Vassilev, T., Verrall, H., Yang, B., Zhao, Z.: Solution of an outstanding conjecture: the non-existence of universal cycles with $k = n - 2$. Discrete Math. **258**, 193–204 (2002)
51. Stinson, D.R.: Combinatorial Designs: constructions and Analysis. Springer, New York (2004)
52. Stinson, D.R., Wei, R., Yin, J.: Packings, chap. VI.40, pp. 550–556. In: Colbourn and Dinitz [14] (2007)
53. Vickers, V.E., Silverman, J.: A technique for generating specialized Gray codes. IEEE Trans. Comput. **29**(4), 329–331 (1980)
54. Wagner, D.G., West, J.: Construction of uniform Gray codes. Congr. Numer. **80**, 217–223 (1991)
55. Wallis, W.D.: Combinatorial Designs. Marcel Dekker, Inc., New York (1988)
56. West, D.B.: Introduction to Graph Theory, second edn. Prentice Hall, Upper Saddle River, NJ (2001)
57. Wilf, H.S.: Combinatorial algorithms: an update. CBMS-NSF Regional Conference Series in Applied Mathematics, 55. Society for Industrial and Applied Mathematics (SIAM), Philadelphia, PA (1989)
58. Williams, A.: Shift Gray codes. Ph.D. thesis, University of Victoria, Victoria, BC (2009)

Chapter 3
Ordering the Blocks of Designs

Just as there are many ways to order subsets, partitions and permutations, there are many ways to order the blocks of a design. A listing of the blocks of a design such that there is a minimal change between consecutive blocks is a combinatorial Gray code. Note that the minimal change definition will depend on the design parameters. For example, the blocks of a $\mathrm{TTS}(v)$ can theoretically be listed such that consecutive blocks differ in exactly one element. On the other hand, for a design with $\lambda = 1$, the strongest minimal change listing would have exactly one element remaining the same from one block to the next. In the previous chapter, we saw that minimal change can also be defined structurally. Ucycles are sequences that adhere to such a definition, although the contents of consecutive objects may also fit the Gray code concept of minimal change. Another rule for ordering blocks of designs that relates both to content and structure employs configurations. Unlike Gray codes and Ucycles, configuration ordering is an ordering paradigm unique to designs.

We begin with configuration orderings. Cohen and Colbourn defined (standard) configuration ordering in [3]. It is a natural idea with practical applications (see Chap. 6). A generalization of this definition was introduced by Dewar in [4]. In addition to work directly on configuration ordering, results in other areas can be naturally presented in these terms. In the second section of this chapter we look at minimal change orderings. First, we look at single-change designs which are objects that by their very definition have a minimal change ordering of their blocks. Next, we look at Gray codes and Ucycles. The definitions we have seen for Gray codes and Ucycles can be naturally extended to blocks of designs. We close the chapter with a brief discussion of other ordering paradigms.

3.1 Configuration Ordering

One way to describe an ordering of the blocks of a design is to describe the intersection pattern induced by (a fixed number of) consecutive blocks. That is, every set of x consecutive blocks should have some common property. This intersection pattern is

M. Dewar and B. Stevens, *Ordering Block Designs: Gray Codes, Universal Cycles and Configuration Orderings*, CMS Books in Mathematics,
DOI 10.1007/978-1-4614-4325-4_3, © Springer Science+Business Media New York 2012

best described by configurations. We begin with the most basic requirement for such an ordering: that every set of x consecutive blocks forms the same configuration. This definition is then generalized to allow every set of x consecutive blocks to form a configuration isomorphic to one of some specified set of allowable configurations. Finally, we discuss decompositions where the patterns formed are not necessarily configurations in the strict design theoretic definition; however, they naturally fit under this rubric.

3.1.1 Standard Configuration Orderings

The majority of research regarding configurations has focused on the presence or absence and the enumeration of given configurations in certain families of designs. For example, anti-Pasch triple system existence has been settled. The other most commonly studied aspect of configurations is the question of whether or not a given design can be decomposed into copies of a given configuration. In 2003, another area of investigation regarding configurations was identified [3]. Cohen and Colbourn's configuration ordering is a listing of the blocks of a design such that every set of consecutive blocks of the necessary number is isomorphic to a given configuration. The requirements for existence of a configuration ordering are stronger than for existence of a decomposition of a design into copies of the same configuration. That is, any configuration ordering for a design gives a decomposition of that design into copies of the configuration in question. The main question asked is, given a configuration, does a design admit a configuration ordering of its blocks? Recall that a (p,ℓ)-configuration is a set system with p elements and ℓ blocks in which every element is contained in at least one block. Let C be a configuration. In [3], Cohen and Colbourn ask: When does there exist a Steiner triple system of order v in which the triples can be ordered such that every ℓ consecutive triples form a configuration isomorphic to C?, Such an ordering is called a C-ordering for the STS.

Definition 3.1 (configuration ordering). Let $S=(V,\mathscr{B})$ be a BIBD with $|\mathscr{B}|=b$. Let C be a configuration on ℓ blocks. A C-ordering for S is a list of the blocks of S, $B_0,B_1,\dots,B_{b-1}$, with the property that $B_i,B_{i+1},\dots,B_{i+\ell-1}\cong C$, for all $0\le i\le b-\ell$. If $B_i,B_{i+1},\dots,B_{i+\ell-1}\cong C$ holds for all $0\le i\le b-1$, with subscript addition performed modulo b, then the ordering is called ***C*-cyclic**.

For example, the blocks of the Fano plane can be listed such that consecutive sets of three blocks form a triangle or B_5-configuration:

$$\{0,1,3\},\{1,2,4\},\{1,5,6\},\{0,2,6\},\{2,3,5\},\{3,4,6\},\{0,4,5\}.$$

This ordering is cyclic. We will say that a design admitting such an ordering is ***C*-cyclic orderable**.

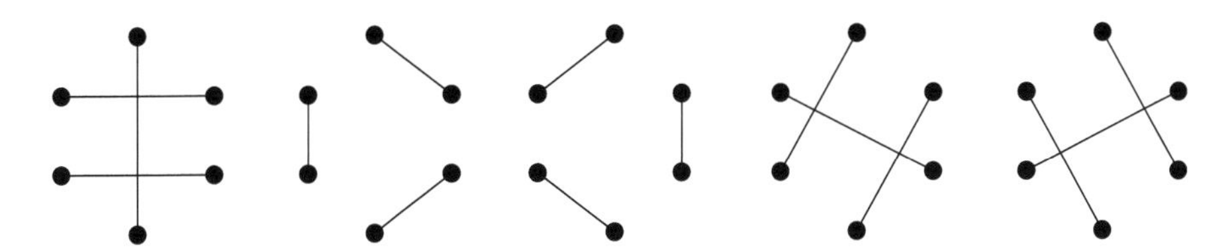

Fig. 3.1 A decomposition of K_6 into 1-factors ordered such that consecutive 1-factors form a Hamilton cycle

3.1.2 Generalized Configuration Orderings

Dewar defined a natural generalization of configuration ordering [4].

Definition 3.2 ($\mathscr{C}$-ordering). Let $S = (V, \mathscr{B})$ be a BIBD with $|\mathscr{B}| = b$. Let $\mathscr{C}$ be a set of configurations, each having ℓ blocks. A $\mathscr{C}$-ordering for S is a list $B_0, B_1, \ldots, B_{b-1}$ with the property that for all $0 \le i \le b-\ell$ there exists $C_j \in \mathscr{C}$ such that $B_i, B_{i+1}, \ldots, B_{i+\ell-1} \cong C_j$. If $B_i, B_{i+1}, \ldots, B_{i+\ell-1} \cong C_j$ holds for all $0 \le i \le b-1$, with subscript addition performed modulo b, then the ordering is called **$\mathscr{C}$-cyclic**.

We will say that a design admitting at least one such ordering is **$\mathscr{C}$-cyclic orderable**.

For example, the following is a $\{(3,3),(4,3)\}$-cyclic ordering for the 2-(5,2,1) design formed by the edges of the complete graph K_5:

$$\{0,1\},\{0,2\},\{1,2\},\{1,3\},\{2,3\},\{2,4\},\{3,4\},\{3,0\},\{4,0\},\{4,1\}.$$

Many results in other areas can be expressed within this framework. In Chap. 4 we will discuss results in configuration ordering, both direct and translated. Further, orderings of graph decompositions can also be discussed in this language, and these are addressed in Sect. 4.3. Specific configuration requirements often arise from applications; these will be further discussed in Chap. 6.

3.1.3 Decompositions

Configuration orderings can be extended to decompositions, defined in Sect. 2.2.7, by restricting the possible graphs formed by multiple consecutive copies of the decomposition graph H. In Fig. 3.1 we present a decomposition of K_6 into five 1-factors and an ordering of these such that the union of any consecutive pair forms a Hamilton cycle. In fact, in this small example, it is the case that the union of *any* pair of 1-factors forms a Hamilton cycle; this is not generally the case in larger K_v.

3.2 Minimal Change Ordering

Although the generalized configuration ordering framework is general and flexible enough to encompass every possible ordering of the blocks of a design that we have encountered, it can become awkward and lose its descriptive power when the set of configurations becomes very large. Hence, it is important to discuss some specific ordering frameworks (namely, the minimal change ordering for the blocks of a design) in their own language.

3.2.1 Single-Change Designs

Single-change covering designs were first introduced in [10]. These are a type of design in which ordering the blocks is inherent in their definition. A series of papers were published in the 1990s regarding the existence of single-change designs.

Definition 3.3 (single-change covering design). A single-change covering design (SCCD) based on the set $[v] = \{1,\ldots,v\}$ with block size k is an ordered collection of b blocks, $\mathscr{B} = B_1,\ldots,B_b$, each $B_i \subset [v]$, which obey: (1) each block differs from the previous block by a single element, and (2) every (unordered) pair $\{x,y\}$ of $[v]$, with $x \neq y$, can be written as $\{e_i,z\}$, where $z \in B_i$ and $e_i \in B_i \setminus B_{i-1}$ for some $i = 2,\ldots,b$.

Note that we are dealing with covering designs here. That is, every pair of points from $[v]$ will appear *at least* once in the blocks of the design. The aim is to minimize b given a fixed v and k. A **single-change circular covering design** (SCCCD) is an SCCD with the requirement that property (1) holds from the last block to the first and with the addition that pairs considered to be covered in the first block are based on the element introduced from last to first blocks; that is, $e_1 \in B_1 \setminus B_b$. This differs from non-cyclic SCCDs where the first block is considered to cover all pairs of points in it. The element e_i is said to be **introduced** in block B_i, and the pairs $\{e_i,z\}$, where $z \in B_i$, are **covered** by B_i. It is interesting to note that the order of blocks in an SCCD or an SCCCD can be reversed and the definitions still hold; the proof is left as an exercise.

Let $b(v,k)$ denote the number of blocks in an SCCD(v,k) and let $b_*(v,k)$ denote the minimum number of blocks possible for the given parameters.

Lemma 3.1 ([10]). *The minimum number of blocks in an SCCD(v,k) is*

$$b_*(v,k) \geq \left\lceil \frac{\binom{v}{2} - \binom{k-1}{2}}{k-1} \right\rceil.$$

Proof. Suppose the blocks $B_1,B_2,\ldots,B_i$ of an SCCD are known. The addition of a block B_{i+1} can result in the covering of at most $k-1$ new pairs: if $B_i = \{a_1,a_2,\ldots,a_k\}$ and $B_{i+1} = \{b_1,a_2,\ldots,a_k\}$, the only possible new pairs are

$\{b_1, a_j\}$, for $2 \leq j \leq k$. There are $\binom{v}{2}$ pairs to be covered, and the first block covers $\binom{k}{2}$, while subsequent blocks cover at most $k-1$ pairs; thus,

$$(b-1)\cdot(k-1) \geq \binom{v}{2} - \binom{k}{2}.$$

Rearranging yields

$$b \geq \frac{\binom{v}{2} - \binom{k}{2} + (k-1)}{k-1}$$

which can be simplified to

$$b \geq \left\lceil \frac{\binom{v}{2} - \binom{k-1}{2}}{k-1} \right\rceil$$

with no loss of accuracy. □

An SCCD is said to be **economical** if it attains the equation above with equality. An economical SCCD is **tight** if $k-1$ divides $\binom{v}{2} - \binom{k-1}{2}$ exactly, that is, each pair is covered exactly once. The fact that each pair is covered exactly once does not mean the resulting design is actually a BIBD due to the notion of covering in this context, that is, pairs covered by a block are only those of the form $\{$*new element, old element*$\}$.

The examples in Tables 3.1 and 3.2 are given by McSorley in [8].

McSorley presents lower bounds on the number of blocks for SCCCDs in [8].

Lemma 3.2. *For $v \geq 4$ and $k \geq 3$, the minimum number of blocks in an SCCCD(v,k), satisfies*

$$b_*(v,k) \geq max\left\{v-1, \left\lceil \frac{v(v-1)}{2(k-1)} \right\rceil \right\}.$$

Proof. To see that $b_*(v,k)$ is at least $v-1$, note that exactly one element is introduced per block of the design; therefore, if the number of blocks is strictly less than $v-1$, there are at most $v-2$ elements introduced. There are at least two elements which are never introduced, and this pair is never covered.

The other bound is obtained by the same argument as in the proof of Lemma 3.1: there are $\binom{v}{2}$ pairs to be covered, and each block covers at most $k-1$ pairs; therefore, $b_*(v,k)\cdot(k-1) \geq v(v-1)/2$. □

The same definitions of economical and tight apply to SCCCDs.

An SCCCD(v,k) is in **standardized form** if:

1. The elements of the first block are $1,2,\ldots,k$, in that order.
2. The other elements are introduced initially in the order $k+1, k+2, \ldots, v$.

Table 3.1 An economical SCCCD(6,3)

Blocks	Element introduced (e_i)	Pairs covered ($\{e_i,z\}, z \in B_i$)
$\{6,4,2\}$	6	$\{6,4\},\{6,2\}$
$\{6,3,2\}$	3	$\{3,6\},\{3,2\}$
$\{6,3,5\}$	5	$\{5,6\},\{5,3\}$
$\{6,3,1\}$	1	$\{1,6\},\{1,3\}$
$\{4,3,1\}$	4	$\{4,3\},\{4,1\}$
$\{4,5,1\}$	5	$\{5,4\},\{5,1\}$
$\{2,5,1\}$	2	$\{2,5\},\{2,1\}$
$\{2,4,1\}$	4	$\{4,2\},\{4,1\}$

Table 3.2 A tight SCCCD(5,3)

Blocks	Element introduced (e_i)	Pairs covered ($\{e_i,z\}, z \in B_i$)
$\{1,2,3\}$	3	$\{3,1\},\{3,2\}$
$\{2,3,4\}$	4	$\{4,2\},\{4,3\}$
$\{3,4,5\}$	5	$\{5,3\},\{5,4\}$
$\{4,5,1\}$	1	$\{1,4\},\{1,5\}$
$\{5,1,2\}$	2	$\{2,5\},\{2,1\}$

Table 3.3 A standardized SCCCD(8,3)

1	1	1	1	7	7	7	7	8	6	6	6	1	1
2	2	5	6	6	8	3	3	3	3	2	2	2	2
3	4	4	4	4	4	4	5	5	5	5	8	8	7

3. The elements of the first block are changed initially in the order $k, k-1, \dots, 2, 1$ (if our SCCCD has one element, say element 1, in every block, then the elements of the first block are changed initially in the order $k, k-1, \dots, 2$).
4. Beginning at the second block, a block's unchanged elements are in the same positions as in the previous block (see Table 3.1). This is called column-strict as these designs are often presented in tabular format.

Table 3.3 represents an SCCCD in standardized form. Columns represent blocks.

Definition 3.4 (single-change neighbour design). A single-change neighbour design, SCND(v,k), is an ordered list of cycles $C_1, C_2, \dots, C_b$ of length k chosen from a complete graph K_v, with the following properties: (1) every edge of K_v occurs in at least one of the cycles, and (2) for $1 \le i < b$, C_{i+1} can be obtained from C_i by replacing one vertex with a different vertex of K_v.

Note that the above definition yields an SCCD if the word cycle is changed to clique. Thus, for $k = 3$, SCCDs and SCNDs are equivalent. The minimum number of blocks in an SCND(v,k) is

$$\frac{v(v-1)-2k+4}{4}$$

due to the fact that given a sequence of blocks in an SCND, $C_1, C_2, \ldots, C_i$, the addition of C_{i+1} covers at most two additional edges and the first block of the design covers k edges.

Wallis subsequently defined a more restricted object which is, although non-cyclic, otherwise equivalent to a rank k Ucycle for the blocks of a covering design on v points with block size k [11].

Definition 3.5 (sequential covering design). A sequential covering design, SD (v,k,t), is a sequence of length t of elements from a v-set, V, with the property that every pair of elements from V occur at most $k-1$ positions apart somewhere in the sequence.

We write $g(v,k)$ for the minimum t for which such a sequence exists. Let $f(v,k)$ denote the smallest number of blocks for which an SCCD(v,k) exists. Since the sequence of sets of k consecutive points from an SD(v,k,t) forms an SCCD(v,k) with $t-k+1$ blocks, we know $g(v,k) \geq f(v,k)+k-1$, and thus lower bounds for SCCDs apply to SDs as well. Furthermore, Wallis was able to establish additional lower bounds by counting appearances of a symbol in the sequence, similar to the approach used in Ucycles.

Theorem 3.1 ([11]). *If* $(2s-1)(k-1) < v-1 \leq 2s(k-1)$ *for some s, then*

$$g(v,k) \geq v\left\lceil \frac{v-1}{2k-2} \right\rceil + 1;$$

otherwise,

$$g(v,k) \geq v\left\lceil \frac{v-1}{2k-2} \right\rceil.$$

For example, 512634567853714827 is an SD$(8,3,18)$.

3.2.2 *Gray Codes for Block Designs*

A Gray code for a BIBD(v,k,λ) is similar to a Gray code for k-subsets of an n-set. In this monograph, we deal only with designs where the elements in a block are not considered to be themselves ordered; therefore, we need only consider the most basic definition of a Gray code for k-subsets of an n-set. A list containing all k-subsets of an n-set is said to be in minimal change order if consecutive subsets differ by the deletion of one element and the insertion of another. Such an ordering is not always possible for the blocks of a design. What is minimal change depends on the size of the blocks in question and the strength and index of the design.

Definition 3.6 (κ-intersecting Gray code). A κ-intersecting Gray code for a design is a listing of the blocks of the design such that consecutive blocks intersect in exactly κ points.

Table 3.4 A 2-intersecting cyclic Gray code for a TTS(7) (read column-wise)

$\{0,1,3\}$	$\{2,4,5\}$	$\{3,4,6\}$	$\{0,1,5\}$	$\{0,2,6\}$
$\{1,3,4\}$	$\{2,3,5\}$	$\{0,4,6\}$	$\{1,5,6\}$	$\{0,2,3\}$
$\{1,2,4\}$	$\{3,5,6\}$	$\{0,4,5\}$	$\{1,2,6\}$	

When the required minimal change property holds between the first and last blocks of a Gray code, we refer to it as a **cyclic Gray code**. The intersection value (κ) of a Gray code is related to the strength, block size, and index of the design in question. For example, a TTS(v) may admit a 2-intersecting Gray code for its blocks because every pair of points appears in two blocks. However, an STS(v) cannot possibly admit a 2-intersecting Gray code, since every pair of points appears in a single block, but it may admit a 1-intersecting Gray code. More generally, the blocks of a BIBD($v,k,1$) intersect in at most one point; therefore, a Gray code for these blocks would have consecutive blocks intersecting in exactly one point regardless of the value of k.

We are not aware of any results directly regarding Gray codes for the blocks of designs until the thesis of Dewar [4]. However, a single-change covering design with blocks of size k is a $(k-1)$-intersecting Gray code for the design (which is a covering). The revolving door algorithm (described in Sect. 2.3.2) implies the existence of a 1-intersecting cyclic Gray code for each BIBD($v,2,1$), as these designs are exactly the collection of all 2-subsets of a v-set. The existence of a Hamilton cycle in the 1 block-intersection graph of all BIBD($v,k,1$)s [5] implies the existence of 1-intersecting cyclic Gray codes for the blocks of these designs. In general, the existence of κ-intersecting cyclic Gray codes for BIBD(v,k,λ)s, where $1 \le \kappa \le k-1$, can be established by determining if the κ-block-intersection graph of the BIBD is Hamiltonian. On the other hand, a Ucycle of rank k for the blocks of a BIBD(v,k,λ) is also a $(k-1)$-intersecting cyclic Gray code for the design. We define Ucycles for block designs in the next section.

Table 3.4 is a 2-intersecting cyclic Gray code for a TTS(7).

3.2.3 Universal Cycles for Block Designs

As evidenced by the discussion in Sect. 2.3.4, it is first necessary to fix a formal definition of a Ucycle for a block design. In particular, what (size of) alphabet will be used in the Ucycle and how will blocks be represented by this alphabet? One possible representation of a Ucycle for a BIBD(v,k,λ) is a sequence of elements such that every length k-subsequence is a unique block of the design and such that every block of the design is represented exactly once. We rephrase this definition in the formal language of Chung et al. Let $(V,\mathscr{B})$ be a BIBD(v,k,λ). Let $b = |\mathscr{B}|$ and assume each $B \in \mathscr{B}$ is specified by a sequence $x_0,x_1,\ldots,x_{k-1}$, where, for

$0 \le i \le k-1$, $x_i \in B$. The sequence $\mathbf{U} = a_0, a_1, \ldots, a_{b-1}$ is a Ucycle for $(V, \mathscr{B})$ if $a_i, a_{i+1}, \ldots, a_{i+k-1}$, $0 \le i \le b-1$, runs through each element of $\mathscr{B}$ exactly once, where index addition is performed modulo b. Ucycles of this type exist for all BIBD$(v,2,1)$s with odd v. This is proved by Jackson in [14], although he refers to the collection of blocks as the set of 2-combinations of $[v]$. However, in general, there are three problems with this definition. First, we have restricted ourselves to balanced incomplete block designs, and there is no obvious way to extend the definition to include pairwise balanced designs. Second, the definition fails to work for many BIBDs because it requires $\lambda \ge k-1$. Third, we will run into trouble with higher strength designs. Given a t-(v,k,λ) design and a sequence of points from this design, every set of t consecutive points will appear in $k-t+1$ blocks. As a result, we must have $\lambda \ge k-t+1$. In fact, every set of s consecutive points, for $1 \le s \le t$, will appear in $k-s+1$ blocks. So we must have $\lambda_s \ge k-s+1$, for all $1 \le s \le t$, where

$$\lambda_s = \lambda_t \frac{\binom{v-s}{t-s}}{\binom{k-s}{t-s}}$$

and $\lambda_t = \lambda$.

Consider the Fano plane pictured in Fig. 2.2 (page 17). Any pair of points defines a unique block of this design; hence, starting the Ucycle with the sequence 013 means we cannot add another element to the sequence as no block other than $\{0,1,3\}$ contains 1 and 3. One resolution is to define a Ucycle for the Fano plane (or any other BIBD$(v,k,1)$ or PBD$(v,K,1)$) as a list of elements from V such that every length two subsequence represents a unique block and such that each block is represented exactly once. Under this definition, a Ucycle for the Fano plane in Fig. 2.2 is 3150642. This represents the blocks in the order: $\{0,1,3\}, \{1,5,6\}, \{0,4,5\}, \{0,2,6\}, \{3,4,6\}$, $\{1,2,4\}, \{2,3,5\}$. As the pairwise balanced design obtained by deleting a single point (e.g., the point 6 in Fig. 2.2) from the Fano plane does not have a Ucycle under this definition, we consider this to be a non-trivial definition of a Ucycle for a PBD. This is a natural representation for objects in a Ucycle, and similar representations have been used for other combinatorial objects. For example, Jackson and Buhler have looked at Ucycles for all subspaces of dimension x, $x \ge 2$, in a vector space of dimension greater than x. For their representation of a subspace, they use a set of vectors from a basis of the subspace [2, 7].

A single definition covers both types of Ucycles for PBDs discussed above.

Definition 3.7 (Ucycle of rank κ). Let $(V, \mathscr{B})$ be a PBD(v,K,λ). Let $b = |\mathscr{B}|$ and let each $B \in \mathscr{B}$ be represented by the sequence $x_0, x_1, \ldots, x_{\kappa-1}$, where, for $0 \le i \le \kappa-1$, $x_i \in B$. The sequence $\mathbf{U} = a_0, a_1, \ldots, a_{b-1}$ is a Ucycle of rank κ for $(V, \mathscr{B})$ if $a_i, a_{i+1}, a_{i+2}, \ldots a_{i+\kappa-1}$, $0 \le i \le b-1$, runs through the representation of each element of $\mathscr{B}$ exactly once, where index addition is performed modulo b.

The key point here is that blocks can be represented by objects of various ranks, although all blocks within a given Ucycle must have representations of the same rank. The obvious choices for rank are strength and block size. For example, when representing the whole block of a TTS(v), we would use an object of rank three. However, it is not always possible to use k-subsets to represent blocks of size k. For example, a 2-(v,k,λ) design with $\lambda < k-1$ does not admit a Ucycle where the representative of each block is the block itself. In this case, we represent a block by a pair of points appearing in that block, that is, by an object of rank two which is the strength of the design. Before stating explicit definitions for rank two and rank three Ucycles, we state a necessary condition for the existence of a rank κ Ucycle. Recall that r_x is the number of blocks in which the point x appears.

Lemma 3.3. *If there exists a rank κ Ucycle, U, for the design $S = (V, \mathscr{B})$, then each $x \in V$ appears at most $\lfloor r_x/\kappa \rfloor$ times in U which implies $\sum_{x \in V} \lfloor r_x/\kappa \rfloor \geq b$.*

Proof. Suppose $U = a_0, a_1, \ldots, a_{b-1}$ is a Ucycle of rank κ representing the blocks of S. Consider a fixed symbol $a_i = x$ in U. The symbols a_{i+j}, for $-\kappa < j < \kappa$, must not be equal to x; therefore, each copy of x in U occurs in exactly κ sets of κ consecutive points in U. Each of these κ consecutive point sets represents a block of the design containing the point x. Let n_x represent the number of times x appears in U. The point x appears in exactly r_x blocks of the design; however, a block containing x may not be represented in U by a set including x. Therefore, $n_x \leq \lfloor r_x/\kappa \rfloor$. Since U is of length b, $\sum_{x \in V} n_x = b$, and we have $\sum_{x \in V} \lfloor r_x/\kappa \rfloor \geq b$. □

Lemma 3.3 can be generalized as follows. Let $\mathscr{S}_s^v$ represent the collection of s-subsets of a v-set. For a given design $S = (V, \mathscr{B})$, let A be a subset of V and let r_A denote the number of blocks of the design that the subset A appears in. If $s \leq t$, where t is the strength of S, an s-set A appears in exactly

$$r_A = \frac{\lambda_t \binom{v-s}{t-s}}{\binom{k-s}{t-s}}$$

blocks of the design.

Lemma 3.4. *If there exists a rank κ Ucycle for the design $S = (V, \mathscr{B})$, then*

$$\sum_{A \in \mathscr{S}_s^v} \left\lfloor \frac{r_A}{(\kappa - s + 1)} \right\rfloor \geq b,$$

where $v = |V|$ and $s = |A|$.

Proof. The proof proceeds exactly as in the proof of Lemma 3.3 with n_A representing the number of times a subset A appears consecutively in U (regardless of order). Each contiguous set of s elements in U appears in $\kappa - s + 1$ blocks; however, a block containing A may not be represented in U by a κ-set containing A as a contiguous subsequence of s elements. Therefore, $n_A \leq \lfloor r_A/(\kappa - s + 1) \rfloor$. Since U is of length b, $\sum_{A \in \mathscr{S}_s^v} n_A = b$, and we have $\sum_{A \in \mathscr{S}_s^v} \lfloor r_A/(\kappa - s + 1) \rfloor \geq b$. □

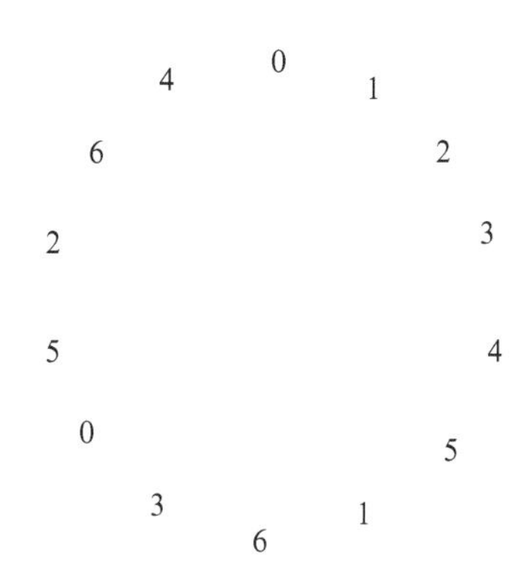

Fig. 3.2 A Ucycle of rank three for the blocks of a TTS(7)

Although Lemmas 3.3 and 3.4 are general necessary conditions for the existence of a rank κ Ucycle, these conditions are weak because of the allowance for variable rank. If the design is regular (i.e., has all blocks the same size) and the rank of the Ucycle is equal to the size of the blocks of the design, we have the following refinement of Lemma 3.3.

Lemma 3.5. *Let $S = (V, \mathscr{B})$ be a BIBD(v,k,λ). If there exists a rank k Ucycle for S, then k divides r, where $r = \lambda(v-1)/(k-1)$.*

Proof. Let U be a rank k Ucycle for S. The rank of U is equal to the block size of S; hence, each block represents itself in U. Consider a fixed symbol a_i in U and suppose $a_i = x$. The symbols a_{i+j}, for $-k < j < k$, must not be equal to x; therefore, each copy of x in U occurs in exactly k k-subsequences of consecutive points in U. Each k-subsequence represents a block of S containing x; there are $r = \lambda(v-1)/(k-1)$ such blocks. Therefore, x must appear r/k times in U. As this value must be an integer, we must have $k|r$. □

In this monograph, the only Ucycles with rank equal to block size that we consider are for triple systems; therefore, we restrict the definition of a rank three Ucycle to the following.

Definition 3.8 (Ucycle of rank three). Let $(V, \mathscr{B})$ be a TS(v,λ), with $\lambda \geq 2$. Let $b = |\mathscr{B}|$ and let each block $B \in \mathscr{B}$ be represented by the sequence x_0, x_1, x_2, where $x_i \in B$ for $0 \leq i \leq 2$. The sequence $\mathbf{U} = a_0, a_1, \ldots, a_{b-1}$ is a Ucycle for the TS(v,λ) if a_i, a_{i+1}, a_{i+2}, $0 \leq i \leq b-1$, runs through each element of $\mathscr{B}$ exactly once, where index addition is performed modulo b.

Figure 3.2 represents a Ucycle of rank three for a TTS(7).

For triple systems with $\lambda = 1$, and for all other PBDs, we will look at the existence of rank two Ucycles. In these cases, the rank is equal to the strength of the design.

Definition 3.9 (Ucycle of rank two). Let $(V, \mathscr{B})$ be a PBD(v,K,λ). Let $b = |\mathscr{B}|$ and let each $B \in \mathscr{B}$ be represented by a sequence x_0, x_1, where, for $i = 0, 1$, $x_i \in B$.

The sequence $\mathbf{U} = a_0, a_1, \ldots, a_{b-1}$ is a Ucycle for the PBD(v, K, λ) if a_i, a_{i+1}, $0 \leq i \leq b-1$, runs through the representation of each element of $\mathscr{B}$ exactly once, where index addition is performed modulo b.

Notice that in this definition the representation of blocks is not unique. That is, there are $\binom{k}{2}$ ways to represent a block of size k. In fact, taking order into account means that there are $2\binom{k}{2}$ ways to represent a block of size k. Furthermore, when $\lambda > 1$, x_0, x_1 will appear in λ different blocks; therefore, each pair of consecutive points in the Ucycle does not represent a unique block. We will address these issues in Sect. 5.3.

3.3 Other Ordering Paradigms

The primary focus of this monograph is to survey and study orderings that can be expressed as configuration orderings. We think of this as a local constraint on ordering the blocks of a design. We strongly suspect that all local notions of ordering can be put into configuration ordering terminology, albeit possibly awkwardly. However, these are not the only kinds of ordering that have been studied in combinatorial design theory. There are two other forms of ordering where the constraint is at a global, rather than local, level. Although they are not our primary focus, we want this book to be a comprehensive source for anyone interested in ordering the blocks of designs; therefore, we define these other orderings, and in Chap. 6 we briefly survey the known material in these research areas.

3.3.1 Induced Set Systems

In applications to tournament scheduling and to group testing, we will see orderings of blocks where what defines the ordering is not that each consecutive set of blocks is isomorphic to some configuration. Rather, each consecutive set of blocks determine some object, which may be as simple as the union or more structured. The requirement on the ordering is that the whole set of these induced objects be a set system with some property. In tournament scheduling, each consecutive pair of matches played by a single team induces an ordered pair of the two opponents. The collection of these induced pairs must have no repetitions. This global condition is discussed in Sect. 6.2.4.2. In group testing, the relevant objects are the unions of consecutive blocks from the ordering. Typically, they must be distinct, of a fixed size or parity, or some combination of these properties. In fact, in group testing applications, the ordering must satisfy local, configuration-like conditions, and these global conditions too. Group testing is discussed in Sect. 6.4.

3.3.2 Maximizing Objective Functions

Another idea that leads to useful and interesting notions of ordering the blocks of a design is the idea of a prioritized ordering. Following Bryce and Colbourn [1], we adapt a definition put forward by Rothermel et al. [9] to the general setting of block designs.

Definition 3.10 (f-prioritized ordering). Let $(V,\mathscr{B})$ be a block design. Let $P\mathscr{B}$ be the set of all permutations of the block set and let f be a function from $P\mathscr{B} \to \mathbb{R}$. An f-prioritized ordering for $(V,\mathscr{B})$ is a $\pi \in P\mathscr{B}$ such that $f(\pi) \geq f(\pi')$ for all other $\pi' \in P\mathscr{B}$.

Mathematically, if one knows the function f, then there is nothing sophisticated to do other than extract the maximal element from $P\mathscr{B}$. Most research on these orderings concentrates on showing that the function is relevant to the problem at hand or concentrates on finding efficient algorithms for calculating it (or approximations to it). Finding an f-prioritized ordering is often considered to be NP-hard due to the combinatorial explosion of the set $P\mathscr{B}$ and, in some cases, finding an f-prioritized ordering can be shown to be equivalent to the halting problem [9]. As all instances of this notion of ordering occur in specific applications, these will be discussed exclusively in Sect. 6.3. Note that the calculation of priorities may be dynamic and may change based on the partial ordering established thus far. This, and the potential difficulty of an exhaustive search, motivates a second definition.

Definition 3.11 (dynamically f-prioritized). Let $(V,\mathscr{B})$ be a block design with $b = |\mathscr{B}|$ and suppose that a partial sequence, $S_i = B_1,\dots,B_i$, $i \leq b$, of the blocks is given. Let f be a function defined on the sets of blocks in and not in the sequence. Let $B \in \mathscr{B} \setminus \cup_{j=1}^{i} B_j$, then $f(S_i,B) \in \mathbb{R}$. An ordering $S = B_1,B_2,\dots,B_b$ of the blocks of $\mathscr{B}$ is dynamically f-prioritized if

$$f(S_i,B_{i+1}) \geq f(S_i,B') \text{ for all } 0 \leq i < b \text{ and } B' \in \mathscr{B} \setminus \cup_{j=1}^{i} B_j.$$

Exercises and Problems

Exercise 3.1. Prove that the ordering of the blocks of a SCCD or SCCCD can be reversed and the single-change covering properties still hold.

Problem 3.1. What other ranks, besides the strength, t, and the block size, k, are natural to use in Ucycles for designs?

References

1. Bryce, R.C., Colbourn, C.J.: Prioritized interaction testing for pairwise coverage with seeding and constraints. Inform. Software Tech. **48**, 960–970 (2006)
2. Buhler, J., Jackson, B.: Private communication (2006)
3. Cohen, M.B., Colbourn, C.J.: Optimal and pessimal orderings of Steiner triple systems in disk arrays. Theoret. Comput. Sci. **297**, 103–117 (2003)
4. Dewar, M.: Gray codes, universal cycles and configuration orderings for block designs. Ph.D. thesis, Carleton University, Ottawa, ON (2007)
5. Horák, P., Rosa, A.: Decomposing Steiner triple systems into small configurations. Ars Combin. **26**, 91–105 (1988)
6. Jackson, B.W.: Universal cycles of k-subsets and k-permutations. Discrete Math. **117**, 141–150 (1993)
7. Jackson, B.W., Buhler, J., Mayer, R.: A recursive construction for universal cycles of 2-subspaces. Discrete Math. **309**(17), 5328–5331 (2009)
8. McSorley, J.P.: Single-change circular covering designs. Discrete Math. **197/198**, 561–588 (1999)
9. Rothermel, G., Untch, R.H., Chu, C., Harrold, M.J.: Prioritizing test cases for regression testing. IEEE Trans. Software Eng. **27**(10), 929–948 (2001)
10. Wallis, W., Yucas, J.L., Zhang, G.H.: Single-change covering designs. Des. Codes Cryptogr. **3**, 9–19 (1992)
11. Wallis, W.D.: Sequential covering designs. In: Graph Theory, Combinatorics, and Algorithms, (Kalamazoo, MI, 1992), Wiley-Interscience Publication, vol. 1–2, pp. 1203–1210. Wiley, New York (1995)

Chapter 4
Results in Configuration Ordering

In this chapter, we look at configuration orderings for block designs. In Sect. 4.1, we discuss standard configuration ordering. Recall that a standard configuration ordering is a listing of the blocks of a design such that every set of consecutive blocks of the necessary number is isomorphic to a given configuration. Cohen and Colbourn initiated the study of configuration ordering, and they have been the main source of results in this area. We discuss the motivation for their investigations in Chap. 6. In Sect. 4.2, we turn to generalized configuration orderings and discuss the existence of configuration orderings in which every consecutive set of the specified number of blocks is isomorphic to a member of a set of allowable configurations. We conclude the chapter with a look at configuration orderings for graph decompositions other than those represented by BIBDs and PBDs.

Before we begin, we reiterate our notation for configurations. As mentioned previously, all published work regarding configuration ordering, and almost all published work regarding the decomposition of designs into configurations, deals with triple systems. We generalize configurations to BIBDs and PBDs. Given a configuration C having blocks of size three, we will denote the configuration of the same form with blocks of size k by C'. In cases where it is not clear how the configuration should be generalized, we will describe the configuration in detail.

4.1 Standard Configuration Ordering

In this section, we survey results in configuration ordering which involve a single configuration. We break the discussion into two sections: (1) disjoint configurations and (2) configurations with at least one pair of blocks intersecting. For disjoint configurations, we proceed by block size. For intersecting configurations we work our way through configurations following the naming convention of [11], except for a digression into what are called cluttered orderings. For each configuration we proceed, starting with BIBDs, from smaller block size to larger, followed by PBDs.

M. Dewar and B. Stevens, *Ordering Block Designs: Gray Codes, Universal Cycles and Configuration Orderings*, CMS Books in Mathematics,
DOI 10.1007/978-1-4614-4325-4_4,

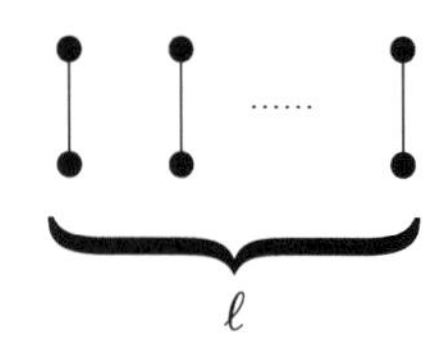

Fig. 4.1 The $(2\ell,\ell)$-configuration (discussed in Theorem 4.1)

4.1.1 Configurations of Disjoint Blocks

We begin with the smallest possible block size—two—but variable number of blocks. This is the $(2\ell,\ell)$-configuration shown in Fig. 4.1.

Theorem 4.1 ([17, 32]). *A BIBD$(v,2,1)$, with $v>4$, admits a $(2\ell,\ell)$-ordering if and only if $\ell \le \lfloor (v-3)/2 \rfloor$.*

We give an outline of the approach of Simmons and Davis to the proof of Theorem 4.1 and encourage the reader to flesh out the details or seek out the original article for further information. The object constructed is called a pair design. Let S_v denote a sequence of adjacent pairs from a length v cycle, starting with $(1,2)$. For v odd, S_v is the length v sequence $(1,2),(3,4),\ldots,(v-2,v-1),(1,v),(2,3),\ldots,(v-1,v)$. For v even, S_v is the length $v/2$ sequence $(1,2),(3,4),\ldots,(v-1,v)$. Let ρ be the permutation $(1)(3\ 5\ 7\ \ldots\ v\ \ldots\ 6\ 4\ 2)$. Simmons and Davis prove that $S_v,\rho S_v,\ldots,\rho^L S_v$ is a pair design, where $L=(v-3)/2$ for v odd and $L=v-2$ for v even.

Pair designs come from the statistical rating technique known as "the method of paired comparisons" which we discuss in Sect. 6.5.1. The $\binom{v}{2}$ pairs from v objects are presented to a subject who must choose between the members of each pair. Studies have shown that if an object is rejected from one pair, there is a bias towards rejecting it again if it appears in a successive pair. Therefore, it is desirable to separate occurrences of an element by as great a distance as possible. Formally, a **pair design** is an ordering of the $\binom{v}{2}$ pairs from v elements so that no two consecutive occurrences of any element have fewer than $\lfloor (v-3)/2 \rfloor$ pairs separating them. Pair designs are also a solution to the problem of scheduling single-match round-robin tournaments when it is desired that each player rest as long as possible between matches. Scheduling tournaments will be discussed in Sect. 6.2.

Let D_ℓ be the $(3\ell,\ell)$-configuration consisting of ℓ disjoint triples. Cohen and Colbourn call a D_ℓ-ordering for an STS a **pessimal ordering**. The next two results were developed in relation to ordering the disks in redundant arrays of independent disk (RAID) systems and are further discussed in the Chap. 6.

Theorem 4.2 ([7]). *For all admissible $v \ge 9\ell-6$, there exists an STS(v) with a D_ℓ-ordering.*

Theorem 4.3 ([7]). *Let ℓ be a positive integer. For all admissible $v \ge 81(\ell-1)+1$, every STS(v) admits a D_ℓ-ordering.*

Proof. Suppose $S = (V, \mathscr{B})$ is an STS(v). Let G be the 1 block-intersection graph of S. G is regular of degree $3(r-1) = 3(v-3)/2$. By Brookes' theorem (Theorem 2.6), G has a proper vertex colouring in $s \leq 3(v-3)/2$ colours. Partition the vertices of G by colour class, letting R_i represent the set of vertices having colour i. Assume $|R_i| \leq |R_{i+1}|$ for all $1 \leq i < s$. If $3|R_1| < |R_s|$, there is a triple in R_s that intersects no triple of R_1. Move that triple from R_s to R_1, relabelling the R_i if necessary to maintain $|R_i| \leq |R_{i+1}|$. Repeat this process until $3|R_1| \geq |R_s|$.

The total number of triples in S is $v(v-1)/6$; therefore,

$$\sum_{i=1}^{s} |R_i| = \frac{v(v-1)}{6}.$$

Since the partitions are ordered by increasing size and R_s is the last partition in the order, we have $|R_s| \geq \lceil v/9 \rceil$. That is, the size of R_s is greater than or equal to the average size of a partition, which is $v(v-1)/(6s)$. Using the relation established between R_1 and R_s, we obtain $|R_1| \geq \lceil v/27 \rceil$. Since $v \geq 81(\ell - 1) + 1$ we have $|R_1| > 3\ell - 3$. The D_ℓ-ordering is obtained by placing all triples of R_i before those of R_{i+1} for all $1 \leq i < s$. Within classes, order the triples as follows: Order the triples of R_1 arbitrarily and suppose that $R_1, \ldots, R_{i-1}$ have been ordered. When selecting the jth triple of R_i it is necessary to ensure that it does not intersect any of the last $\ell - j$ triples of R_{i-1}. Since $j-1$ triples have already been chosen from R_i, and the total number of triples in R_i that may intersect the last $\ell - j$ triples of R_{i-1} is at most $3(\ell - j)$, there are at most $(j-1) + 3(\ell - j) = 3\ell - 2j - 1$ triples of R_i that we cannot use. However, there are at least $3\ell - 2$ triples in R_i, so there exists a triple that can be placed in the jth position in the ordering for R_i. □

When $\ell = 2$, Theorem 4.3 implies that every STS(v), $v \geq 82$, admits an A_1-ordering (see Fig. 1.1 on page 4). In [13] it is shown that every TS(v, λ), $v \geq 17$, admits an A_1-cyclic ordering. The proof is based on the fact that existence of an A_1-cyclic ordering is equivalent to existence of a Hamilton cycle in the complement of the block-intersection graph of the design.

Theorem 4.4 ([13]). *Every TS(v, λ), $v \geq 17$, admits an A_1-cyclic ordering.*

Proof. Let $S = (V, \mathscr{B})$ be a TS(v, λ). The complement of the block-intersection graph of S is the graph with vertex set $\mathscr{B}$ and edge set $\{(B_i, B_j) : B_i \cap B_j = \emptyset, B_i, B_j \in \mathscr{B}\}$. Denote this graph $\overline{G_S}$. A Hamilton cycle in this graph yields an ordering of blocks such that every pair of consecutive blocks is disjoint. Under what conditions does this graph admit a Hamilton cycle? We wish to apply Dirac's theorem (Theorem 2.2); therefore, we must determine the minimum vertex degree in $\overline{G_S}$, denoted $\delta(\overline{G_S})$.

The degree of a vertex w in $\overline{G_S}$ is governed by the number of blocks in $\mathscr{B}$ that intersect the block represented by w. Suppose w represents the block B and let n_1 denote the number of blocks containing exactly one point appearing in B, let n_2 denote the number of blocks containing exactly two points appearing in B, and let n_3 denote the number of blocks containing all three points appearing in B (this count will include B). Since every point of the design appears in r blocks, the points in B appear in $3r$ (non-unique) blocks. As the blocks containing two points that appear in B will be counted twice by this calculation and the blocks containing three points that appear in B will be counted three times by this calculation, we have $3r = n_1 + 2n_2 + 3n_3$. On the other hand, there are three distinct pairs of points that can be made from the points appearing in B, and each of these pairs appears in λ blocks of S. Since a block that is a copy of B accounts for the appearance of three of these pairs, we have $3\lambda = n_2 + 3n_3$. The number of blocks intersecting B is $n_1 + n_2 + n_3 - 1 = 3r - 3\lambda + n_3 - 1$. Since the vertex w representing B is adjacent to the vertices representing blocks that do not intersect B, the minimum degree of w is realized when the number of triples intersecting B is maximized. The maximum value of n_3 is λ; therefore, the maximum number of vertices not adjacent to w is $3r - 2\lambda - 1$. Hence, $\delta(\overline{G_S}) \geq (b-1) - (3r - 2\lambda - 1)$, where $b = v(v-1)\lambda/6$.

In order to employ Dirac's theorem, we must determine when $\delta(\overline{G_S}) \geq b - 3r + 2\lambda \geq b/2$ holds. That is, we wish to determine bounds on λ and v such that the following inequality holds:

$$\frac{v(v-1)\lambda}{6} - \frac{3(v-1)\lambda}{2} + 2\lambda \geq \frac{v(v-1)\lambda}{12}.$$

Simplifying this inequality yields $v^2 - 19v + 42 \geq 0$, which holds for all $v \geq 17$ and arbitrary λ. □

Recall that if C is a configuration involving ℓ triples, then a TS(v,λ), $S = (V, \mathscr{B})$, is said to be decomposable into copies of C if either the set of triples $\mathscr{B}$ or the set of triples $\mathscr{B} \setminus T$, where T is a subset of $\mathscr{B}$ containing at most $\ell - 1$ triples, is decomposable into copies of C. We note that an immediate consequence of Theorem 4.4 is the following corollary, proved for STS(v)s by Horák and Rosa in [20].

Corollary 4.1 ([13]). *Every TS(v,λ), $v \geq 17$, is decomposable into A_1 configurations.*

Momihara and Jimbo independently showed that for any $v \geq 10$, there always exists an optimal $(v,3,1)$ packing having an A_1-cyclic ordering, which gives a partial generalization of Theorem 4.2.

Theorem 4.5 ([27]). *For all $v \geq 10$, there exists an optimal $(v,3,1)$ packing having an A_1-cyclic ordering.*

Proof. The proof is constructive. For each equivalence class of $v \pmod 6$, determine base blocks which are developed in some permutation group to generate the design and the ordering. We give one example and invite the reader to seek out further details in the original paper. Let $v = 6m+5$, with m even. The packing is constructed on the point set $\{1,2,\ldots,2m+1\} \times \{1,2,3\} \cup \{\infty_1, \infty_2\}$. First, define a permutation

$$\sigma = (1)(2m+1\ 2m\ \ldots\ 3\ 2),$$

and an operation, $\circ$, on the set $\{1,2,\ldots,2m+1\}$:

$$i \circ j = \begin{cases} \ell & \text{if } i+j \equiv 2\ell \pmod{2m+1} \\ m+\ell+1 & \text{if } i+j \equiv 2\ell+1 \pmod{2m+1}. \end{cases}$$

Next, define some initial sequences of blocks. The notation $(S_i|1 \leq i \leq j)$ will denote the concatenation of the sequences $S_1, S_2, \ldots, S_j$. For $1 \leq i \leq m-1$, let

$$\begin{aligned} S_i = {} & \{\infty_1,(2i+3)_2,(2i+3)_3\},\{\infty_2,2i_2,2i_3\},\{\infty_1,(2i+2)_1,(2i+2)_2\}, \\ & \{\infty_2,(2i+1)_1,(2i+1)_2\},\{\infty_1,2i_1,2i_2\},\{\infty_2,(2i+3)_1,(2i+3)_2\}, \\ & \{\infty_1,(2i+1)_2,(2i+1)_3\},\{\infty_2,(2i+2)_2,(2i+2)_3\},\{\infty_1,2i_3,(2i+1)_1\}, \\ & \{\infty_2,2i_1,\sigma(2i)_3\},\{\infty_1,(2i+2)_3,(2i+3)_1\},\{\infty_2,(2i+2)_1,\sigma(2i+2)_3\}. \end{aligned}$$

For each $1 \leq h,i,j \leq 2m+1$, $h \neq i$, let

$$\begin{aligned} S'_{h,i,\ell} = {} & \{(h+\ell)_1,(i+\ell)_1,((h+\ell)\circ(i+\ell))_2\}, \\ & \{(h+\ell)_2,(i+\ell)_2,((h+\ell)\circ(i+\ell))_3\}, \\ & \{(h+\ell)_3,(i+\ell)_3,((h+\ell)\circ(i+\ell))_1\}, \end{aligned}$$

and let $S_{h,i,j} = (S'_{h,i,\ell}|0 \leq \ell \leq j)$.

An A_1-cyclic ordering for the packing is

$$\begin{aligned} S = {} & \{\infty_2,1_2,1_3\},(S_i|i=1,3,5,\ldots,m-1),\{\infty_1,1_1,1_3\},\{1_2,2_2,(m+2)_3\}, \\ & \{1_1,2_1,(m+2)_2\},\{1_3,2_3,(m+2)_3\},S_{2,3,2m-1},S_{1,3,2m-2},\{2m_1,1_1,(2m+1)_2\}, \\ & \{2m_3,1_3,(2m+1)_1\},\{2m_2,1_2,(2m+1)_3\},\{(2m+1)_2,2_2,1_3\}, \\ & \{(2m+1)_3,2_3,1_1\},\{(2m+1)_1,2_1,1_2\},\{(2m+1)_2,3_2,(m+2)_3\},S_{1,4,2m-1}, \\ & \{(2m+1)_3,3_3,(m+2)_1\},\{(2m+1)_1,3_1,(m+2)_2\},(S_{1,i,2m}|i=5,\ldots,m+1). \end{aligned}$$

□

The application of A_1-orderings to group testing is discussed in Sect. 6.4.

Adding another disjoint block to the configuration, Theorem 4.3 implies that every STS(v), $v \geq 163$, admits a B_1-ordering (see Fig. 1.2 on page 5). A stronger result is proved in [13]. Note that the proof of this result is very similar in flavour to the proofs in Sects 2.3.3 and 2.3.4 which use the de Bruijn graph to prove existence of a de Bruijn sequence and the transition digraph to prove existence of universal cycles (Ucycles) for k-subsets of an n-set and for permutations.

Theorem 4.6 ([13]). *Every TS(v,λ), $v \geq 103$, admits a B_1-cyclic ordering.*

Proof. Let $S = (V, \mathscr{B})$ be a TS(v, λ). Although we can no longer use a block-intersection graph to effectively represent the ordering question, we again cast the problem in graph theoretic terms. We begin by decomposing the blocks of S into pairs of disjoint blocks. Corollary 4.1 states that such a decomposition exists provided that $v \geq 17$. Create a graph where each vertex represents one of the pairs of the decomposition and denote this graph G. If there is a block not paired in the decomposition, create a special vertex containing three disjoint triples. We must prove that such a vertex can be created.

Recall that the existence of an A_1-cyclic ordering for the blocks of a design is determined by the existence of a Hamilton cycle in the complement of the block-intersection graph for the design. To create the special vertex containing three disjoint triples, we find two consecutive blocks in this Hamilton cycle that are disjoint from the triple leftover after decomposition. Suppose the block $B \in \mathscr{B}$ is not paired in the decomposition of S into A_1 configurations. Recall from the proof of Theorem 4.4 that the maximum number of blocks of S that intersect B in at least one point is $3r - 2\lambda - 1$ (not including B itself). Suppose that no two consecutive blocks in the Hamilton cycle do not intersect B. What is the maximum number of blocks that can be in such a Hamilton cycle? To create pairs from the Hamilton cycle, we first remove the leftover block from the cycle, leaving a path where the first and last blocks do not intersect the leftover block. The worst possible arrangement in this path of the $3r - 2\lambda - 1$ blocks that intersect the leftover block is to have one appear every alternate block; such a sequence contains $2(3r - 2\lambda - 1) + 1$ blocks. Therefore, if $6r - 4\lambda - 1 < b - 1$, we will be able to construct the special vertex. In terms of v and λ, this requirement is $(v^2 - 19v + 42)\lambda > 0$. Thus, for all $v \geq 17$ and arbitrary λ, if $|\mathscr{B}| \equiv 1 \pmod 2$, the required special vertex containing three disjoint triples can be formed.

To complete the construction of G, join two vertices of G by an edge if and only if the blocks represented by them are disjoint. We claim that a Hamilton cycle in G will provide a B_1-cyclic ordering for the blocks of S. Figure 4.2 displays a portion of the graph G.

To see that the ordering of vertices given by a Hamilton cycle in G is a B_1-cyclic ordering for the blocks they represent, consider the underlying blocks. Suppose the Hamilton cycle passes through v_0, v_6, v_4, v_{10} (see Fig. 4.2). This segment of the cycle represents the block ordering $B_0, B_1, B_6, B_{12}, B_4, B_5, B_{14}, B_{10}, B_{11}$. Every set of three consecutive blocks in this list forms a B_1 configuration. The order in which we list the blocks represented by a vertex is arbitrary.

To determine when G admits a Hamilton cycle we appeal to Dirac's theorem (Theorem 2.2). In order to determine the smallest possible minimum degree of a vertex in G, we begin by considering the blocks represented by a given vertex and determining the maximum number of blocks of S that intersect these blocks. There are two cases to consider.

Case 1: $|\mathscr{B}| \equiv 0 \pmod 2$. When the number of blocks in the design is even, no special vertex appears in G. Consider any vertex in G and denote this vertex by w. The vertex contains two disjoint blocks and therefore six distinct points of V.

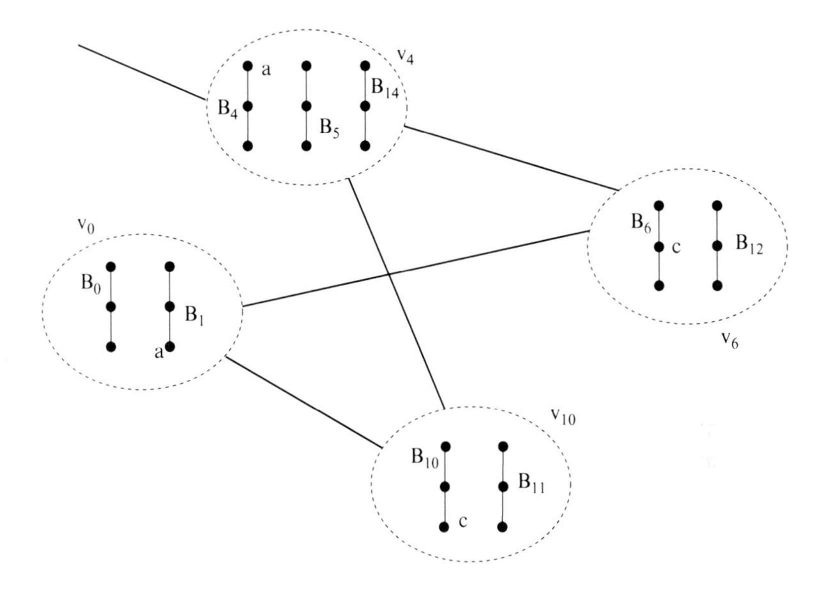

Fig. 4.2 A portion of G (for B_1-cyclic ordering)

Let n_1 denote the number of blocks containing exactly one of these six points, let n_2 denote the number of blocks containing exactly two of these six points, and let n_3 denote the number of blocks containing three of these points (this count will include the two blocks represented by w). Since every point of the design appears in r blocks, the six points represented by w appear in $6r$ (non-unique) blocks. As blocks containing two of these points will be counted twice by this calculation and blocks containing three of these points will be counted three times by this calculation, we have $6r = n_1 + 2n_2 + 3n_3$. On the other hand, there are 15 distinct pairs of points that can be made from the six points represented by w, and each of these pairs appears in λ blocks of the design. Since a block consisting of three of the six points accounts for the appearance of three of these pairs, we have $15\lambda = n_2 + 3n_3$. Finally, note that the two blocks represented by w are included in n_3, so to determine the maximum number of blocks intersecting w, we maximize $n_1 + n_2 + n_3 - 2 = 6r - 15\lambda + n_3 - 2$.

To determine the maximum value of n_3, note that the pairs of points that can be made from the points appearing in w appear 15λ times in the design. Each of the blocks containing three points that appear in w accounts for three of these pairs; therefore, $3n_3 \leq 15\lambda$, which implies that $n_3 \leq 5\lambda$. We conclude that the maximum number of blocks that may intersect the blocks represented by w is $6r - 15\lambda + 5\lambda - 2 = 6r - 10\lambda - 2$. If each of these blocks appears as a member of a different pair of the decomposition (i.e. in a different vertex of G), then w is not adjacent to any of these vertices. The number of vertices in G is $v(v-1)\lambda/12$; therefore, the minimum degree of a vertex in G is $v(v-1)\lambda/12 - 1 - (6r - 10\lambda - 2)$. Hence, $\delta(G) \geq v(v-1)\lambda/12 - 3(v-1)\lambda + 10\lambda + 1$.

When is $\delta(G)$ greater than or equal to $|V(G)|/2$? We must solve the following inequality:

$$\delta(G) \geq \frac{v(v-1)\lambda}{12} - 3(v-1)\lambda + 10\lambda + 1 \geq \frac{v(v-1)\lambda}{24}. \quad (4.1)$$

Simplifying yields

$$v(v-1)\lambda - 72(v-1)\lambda + 240\lambda + 24 \geq 0 \quad (4.2)$$

$$\lambda v^2 - 73\lambda v + 312\lambda + 24 \geq 0. \quad (4.3)$$

All triple systems (with an even number of blocks) for which equation (4.3) holds satisfy the conditions of Dirac's theorem. That is, the graph G induced by the triple system has a Hamilton cycle when $\lambda v^2 - 73\lambda v + 312\lambda + 24 \geq 0$. We conclude that every $\mathrm{TS}(v,\lambda)$, with $v \geq 69$ and an even number of blocks, admits a B_1-cyclic ordering.

Case 2: $|\mathscr{B}| \equiv 1 \pmod 2$. When the number of blocks in S is odd, a special vertex will appear in G. The special vertex has the largest potential number of blocks intersecting it due to the fact that it contains nine distinct points (versus six for all other vertices). Let n_1 denote the number of blocks containing exactly one of the points appearing in the special vertex, let n_2 denote the number of blocks containing exactly two of the points appearing in the special vertex, and let n_3 denote the number of blocks containing three points appearing in the special vertex (this count will include the three blocks represented by the special vertex). Using the same reasoning as in Case 1, we have $9r = n_1 + 2n_2 + 3n_3$ and $36\lambda = n_2 + 3n_3$. Note that the three blocks in the special vertex are included in n_3, so to determine the maximum number of blocks intersecting the blocks of the special vertex, we maximize $n_1 + n_2 + n_3 - 3 = 9r - 36\lambda + n_3 - 3$.

To determine the maximum value of n_3, note that the pairs of points that can be made from the points appearing in the special vertex appear 36λ times in the design. Each of the blocks containing three points that appear in the special vertex accounts for three of these pairs; therefore, $3n_3 \leq 36\lambda$, which implies that $n_3 \leq 12\lambda$. We conclude that the maximum number of blocks that may intersect the blocks of the special vertex is $9r - 36\lambda + 12\lambda - 3 = 9r - 24\lambda - 3$. If each of these blocks appears as a member of a different pair of the decomposition (i.e. in a different vertex of G), then the special vertex is not adjacent to any of these vertices. The number of vertices in G is $(b-1)/2 = (v(v-1)\lambda - 6)/12$; therefore, the minimum degree of the special vertex is $(v(v-1)\lambda - 6)/12 - 1 - (9r - 24\lambda - 3)$. Hence, $\delta(G) \geq (v(v-1)\lambda - 6)/12 - 9(v-1)\lambda/2 + 24\lambda + 2$.

When is $\delta(G)$ greater than or equal to $|V(G)|/2$? We must solve the following inequality:

$$\delta(G) \geq \frac{v(v-1)\lambda - 6}{12} - \frac{9(v-1)\lambda}{2} + 24\lambda + 2 \geq \frac{v(v-1)\lambda - 6}{24}. \quad (4.4)$$

Simplifying yields

$$v(v-1)\lambda - 6 - 108(v-1)\lambda + 576\lambda + 48 \geq 0 \tag{4.5}$$

$$\lambda v^2 - 109\lambda v + 684\lambda + 42 \geq 0. \tag{4.6}$$

All triple systems (with an odd number of blocks) for which equation (4.6) holds satisfy the conditions of Dirac's theorem. That is, the graph G induced by the triple system has a Hamilton cycle when $\lambda v^2 - 109\lambda v + 684\lambda + 42 \geq 0$. We conclude that every $\mathrm{TS}(v,\lambda)$, $v \geq 103$, admits a B_1-cyclic ordering. □

Horák and Rosa have proved that every $\mathrm{STS}(v)$, $v \geq 27$, can be decomposed into B_1 configurations [20]. Theorem 4.6 allows us to generalize this result.

Corollary 4.2. *Every* $TS(v,\lambda)$, $v \geq 69$, *can be decomposed into* B_1 *configurations.*

The careful reader will note that Corollary 4.2 has a different lower bound than Theorem 4.6. We outline the proof of the corollary to indicate how this new lower bound is achieved; however, the details are left as an exercise.

Proof (Proof of Corollary 4.2). Case 1 of the proof of Theorem 4.6 deals with the existence of B_1-cyclic orderings when $b \equiv 0 \pmod 2$, that is, when G does not contain a special vertex. In this case, if $\lambda v^2 - 73\lambda v + 312\lambda + 24 \geq 0$, then S admits a B_1-cyclic ordering. This inequality holds for all $v \geq 69$ and arbitrary λ. Such an ordering implies that the blocks of S can be decomposed into B_1 configurations, with at most two blocks leftover. If $b \equiv 0 \pmod 6$, this decomposition is exact.

When the number of blocks in S is odd, there is a single block, say B, leftover after the decomposition of the design into A_1 configurations (the configurations that make up the vertices of G). Ignore this block, which can be arbitrarily chosen, and proceed to determine the existence of a Hamilton cycle in G. This construction allows us to avoid a special vertex, and applying Dirac's theorem (Theorem 2.2) yields the result that G admits a Hamilton cycle for all $v \geq 69$ and arbitrary λ.

It simply remains to deal with the leftover block. When $b \equiv 1,2 \pmod 3$, the Hamilton cycle can either be exactly partitioned into B_1 configurations or can be partitioned into B_1 configurations with one block remaining; therefore, these triple systems are B_1-decomposable with either one or two blocks remaining. When $b \equiv 0$ (mod 3), the Hamilton cycle can be partitioned into B_1 configurations with two blocks remaining. In this case, there are three blocks of the triple system that are not members of a B_1 configuration; therefore, the triple system has not yet been decomposed into B_1 configurations. Note that the two blocks remaining after the Hamilton cycle have been decomposed into B_1 configurations must form an A_1 configuration. We must ensure that this A_1 configuration is disjoint from the leftover block B. Since B can be arbitrarily chosen, to ensure we have an A_1 configuration that is disjoint from the leftover block, we simply require $b \geq (6r - 10\lambda - 2) + 2 + 1$. Simplifying this inequality yields $\lambda v^2 - 19\lambda v + 78\lambda - 6 \geq 0$. This inequality holds for all $v \geq 17$, so the blocks of any $\mathrm{TS}(v,\lambda)$, $v \geq 17$, can be decomposed into A_1 configurations such that there exists at least one A_1 configuration that is disjoint from the leftover block. □

We now move on to blocks of size k. The existence of A_1'-orderings (i.e. $(2k,2)$-orderings) in the restricted general case of BIBD$(v,k,1)$s is easy to determine using the method employed for triple systems.

Theorem 4.7 ([13]). *Every BIBD$(v,k,1)$, with $v \geq 2k^2+1$, admits an A_1'-cyclic ordering.*

Proof. Let $S=(V,\mathscr{B})$ be a BIBD$(v,k,1)$. We follow the same argument as in the proof of Theorem 4.4; that is, we determine when the complement of the block-intersection graph of S admits a Hamilton cycle. Such a cycle is exactly an A_1'-cyclic ordering since every pair of consecutive blocks is disjoint. We begin by determining the minimum degree of a vertex in the complement of the block-intersection graph. Since $\lambda = 1$, each block of S intersects exactly $k(r-1)$ other blocks, where $r = (v-1)/(k-1)$. This implies that the minimum degree of a vertex in the complement of the block-intersection graph is $(b-1)-kr+k$, where $b = v(v-1)/k(k-1)$. If the minimum vertex degree is equal to or exceeds half the number of vertices in the graph, the graph is Hamiltonian (Theorem 2.2). We must determine when $b-kr+k-1 \geq b/2$. In terms of v and k, this equation is equivalent to $v^2-(2k^2+1)v+2k(k^2-k+1) \geq 0$, which holds whenever

$$v \geq \frac{2k^2+1+\sqrt{4k^4-8k^3+12k^2-8k+1}}{2}. \tag{4.7}$$

Note that $4k^4-8k^3+12k^2-8k+1$ (the quantity under the square root) is less than $4k^4$ for all $k \geq 2$. Therefore, every BIBD$(v,k,1)$, with $v \geq (4k^2+1)/2$, admits an A_1'-cyclic ordering. □

Corollary 4.3 ([13]). *Every BIBD$(v,k,1)$, with $v \geq 2k^2+1$, can be decomposed into A_1' configurations.*

Determining the minimum degree of a vertex in the complement of the block-intersection graph by counting the maximum number of blocks intersecting a given block is difficult for $\lambda > 1$ and general k. A different approach is required to determine the existence of A_1'-orderings for BIBDs in general; however, Momihara and Jimbo have been able to establish the existence of A_1'-cyclic orderings for some Steiner quadruple systems.

Theorem 4.8 ([26]). *There exists an infinite family of SQS(v)s which admit A_1'-cyclic orderings. Small orders in this family include*

$$\begin{aligned} v = {} & 14,16,28,32,40,46,56,64,80,82,92,94,112,118,128,136,160,164,\\ & 166,184,188,190,224,236,238,244,256,272,274,280,320,328,332,\\ & 334,352,368,376,380,382,406,448,472,476,478,488,490,496. \end{aligned}$$

Their construction is recursive and uses the following lemma.

Lemma 4.1 ([26]).

1. *If there exists an* $SQS(v)$ *having an* A_1'*-cyclic ordering, then there exists an* $SQS(2v)$ *having an* A_1'*-cyclic ordering.*
2. *If there exists an* $SQS(v)$ *having an* A_1'*-cyclic ordering, then there exists an* $SQS(3v-2)$ *having an* A_1'*-cyclic ordering.*

These SQSs have application to group testing, discussed in Sect. 6.4.

The method used in proving the existence of B_1-cyclic orderings (see the proof of Theorem 4.6) can be employed to determine the existence of B_1'-orderings (i.e. $(3k,3)$-orderings) for some BIBD(v,k,λ)s. The method requires the existence of an A_1' decomposition. We do not know of any published results regarding the decomposition of BIBD(v,k,λ)s into A_1' configurations, but, Corollary 4.3 implies that all BIBD$(v,k,1)$s, with $v \geq 2k^2+1$, admit decompositions into A_1' configurations.

Theorem 4.9 ([13]). *Every BIBD*$(v,k,1)$*, with* $v \geq 12k^2+1$*, admits a* B_1'*-cyclic ordering.*

Proof. Let $S=(V,\mathscr{B})$ be a BIBD$(v,k,1)$ with $v \geq 2k^2+1$. Corollary 4.3 implies that the blocks of S can be decomposed into A_1' configurations. Create a graph where each vertex represents one of the pairs of the decomposition and denote this graph G. If there is a block not paired in the decomposition, create a special vertex containing three disjoint triples. It is necessary to prove that such a vertex can be created. This is left as an exercise.

To complete the construction of G, join two vertices of G by an edge if and only if the blocks represented by them are disjoint. We claim that a Hamilton cycle in G will provide a B_1'-cyclic ordering for the blocks of S. To determine when G admits a Hamilton cycle, we appeal to Theorem 2.2; therefore, we must determine the maximum number of blocks of S that intersect the blocks of a given vertex. As in the proof of Theorem 4.6, this is broken into two cases based on the parity of the number of blocks in the design. The reader is left to complete the details. □

Corollary 4.4. *Every BIBD*$(v,k,1)$*, with* $v \geq 12k^2+1$*, can be decomposed into* B_1' *configurations.*

4.1.2 Connected Configurations

We now look at configurations in which some blocks intersect. We start with configurations on two blocks which intersect in exactly one point. For triple systems, these are known as A_2 configurations (see Fig. 1.1 on page 4). Beginning with BIBD$(v,2,1)$s, the next result follows from basic graph theory.

Theorem 4.10. *Every BIBD*$(v,2,1)$ *admits a* $(3,2)$*-cyclic ordering.*

Proof. The underlying graph for a BIBD$(v,2,1)$ is K_v. In fact, since the block size is two, each edge of K_v represents a block. The block intersection graph of BIBD$(v,2,1)$ is simply the line graph of K_v and is Hamiltonian (in fact, it is pancyclic [34]). The result follows from the fact that a Hamilton cycle in the line graph is a listing of all edges of K_v (i.e. all blocks of the BIBD$(v,2,1)$) such that consecutive edges share a common vertex. □

In 1989, Ron Graham asked: is the block-intersection graph of an STS(v) Hamiltonian? Several authors have addressed the generalization of this question: is the 1 block-intersection graph of a strength two design Hamiltonian? The answers to these questions lead to several results in configuration ordering because the 1 block-intersection graph of a design having a Hamilton cycle is equivalent to the design admitting a configuration ordering in which every pair of consecutive blocks intersects in exactly one point.

Horák, Pike and Raines proved that the 1 block-intersection graph of every TS(v,λ), with $v \geq 12$ and arbitrary λ, is Hamiltonian [19].

Theorem 4.11 ([19]). *Every TS(v,λ), with $v \geq 12$, admits an A_2-cyclic ordering.*

The proof of Theorem 4.11 employs a theorem of Chvátal and Erdős (Theorem 2.3) which compares the vertex connectivity with the vertex-independence number. This theorem is also used in the proof of Theorem 4.14. We will present the proof of Theorem 4.14 in full, but note that Horák, Pike and Raines use the same but finer arguments in the proof of Theorem 4.11. In particular, the authors show that the vertex-independence number of the 1 block-intersection graph of a TS(v,λ) is at most $\lambda/2v$ and that the size of a cutset in this graph is at least $\lambda/2v$. The paper of Horák, Pike and Raines [19] is very clear and concise and is recommended reading for anyone interested in results on the Hamiltonicity of block-intersection graphs.

Andrew Jesso, working with David Pike and Nabil Shalaby, recently extended this work to larger blocks sizes. In fact, using a strict inequality in Theorem 2.3, they were able to show that, not only are certain 1 block-intersection graphs Hamiltonian, they are actually Hamilton-connected.

Theorem 4.12 ([22,24]). *Every BIBD$(v,4,\lambda)$, with $v \geq 136$, admits an ordering in which every pair of consecutive blocks intersects in exactly one point.*

Even more recently Andrew Jesso proved the following.

Theorem 4.13 ([23]). *Every BIBD$(v,5,\lambda)$, with $v \geq 305$, admits an ordering in which every pair of consecutive blocks intersects in exactly one point.*

Theorems 4.12 and 4.13 employ the same proof methodology as Theorem 4.11. The authors conjecture that results like these hold for any block size; we list these conjectures in Sect. 4.3.3.

The Hamiltonicity of 1 block-intersection graphs was first proved for BIBD $(v,k,1)$s.

Theorem 4.14 ([20]). *Every BIBD$(v,k,1)$ admits a $(2k-1,2)$-cyclic ordering.*

Proof. Let $S = (V, \mathscr{B})$ be a BIBD$(v,k,1)$ and let G_S be its block-intersection graph. We wish to apply the theorem of Chvátal and Erdős (Theorem 2.3). The vertex-independence number of G_S, $\alpha(G_S)$, is bounded above by v/k. Now we must determine a lower bound on the size of a cutset of G_S. Let C be a cutset of G_S and let A be a component of $G_S \setminus C$. Let B represent the remaining components of the disconnected graph, that is, $V(A) \cup V(B) = V(G_S \setminus C)$. Let $W_A = \{x : x \in V, \exists\, b \in A$ such that $x \in b\}$ and $W_B = \{x : x \in V, \exists\, b \in B$ such that $x \in b\}$. Since there is no edge in G_S joining vertices in A to vertices in B, we must have $W_A \cap W_B = \emptyset$. Consider two cases:

Case 1: $W_A \cup W_B \subsetneq V$. Then there exists a point $c \in V$, where $c \notin W_A \cup W_B$. C contains all blocks containing the point c; thus, $|C| \geq (v-1)/(k-1) \geq v/k$, for all $v > k$.

Case 2: $W_A \cup W_B = V$. Then $|W_A| \geq k$ and $|W_B| \geq k$, since both A and B must contain at least one block. As $W_A \cap W_B = \emptyset$, for each pair (x,y), $x \in W_A$ and $y \in W_B$, the 2-subset $\{x,y\}$ must be contained in a block in C. There are exactly $|W_A| \cdot |W_B|$ such pairs; one block in C (a k-set) contains at most $f(k)$ such 2-subsets, where $f(k) = k^2/4$ if k is even, and $f(k) = (k-1)(k+1)/4$ if k is odd. Thus, provided $v \geq 4k/3$,

$$
\begin{aligned}
|C| &\geq \min\{|W_A| \cdot |W_B|/f(k)\} && (4.8)\\
&= \min\{|W_A| \cdot |v - W_A|/f(k)\} && (4.9)\\
&\geq k(v-k)/f(k) && (4.10)\\
&\geq v/k. && (4.11)
\end{aligned}
$$

The vertex-independence number of G_S does not exceed the connectivity of G_S; therefore G_S is Hamiltonian. A Hamilton cycle in the block-intersection graph for a design with $\lambda = 1$ yields an ordering of the blocks where each consecutive pair intersects in a single point. □

Pike has obtained several other results related to the block-intersection graphs of designs. In [25], he and his co-authors prove that the block-intersection graph of every BIBD(v,k,λ) is pancyclic. Note that for $\lambda > 1$, a Hamilton cycle in the block-intersection graph does not necessarily induce a configuration ordering where consecutive blocks intersect in a single point. Depending on λ, the number of intersections may be large and are not guaranteed to be consistent. Therefore, this result can only be expressed as a generalized configuration ordering. Subsequently, Case and Pike [5] generalized this result, showing that all PBD(v,K,λ)s with arbitrary index $\lambda \geq 1$ and $\max(K) \leq \lambda \min(K)$ have block-intersection graphs that are edge-pancyclic. Again, for $\lambda > 1$, this result can only be expressed in terms of generalized configuration ordering. When $\lambda = 1$, the restriction implies that $K = \{k\}$, and thus we have a generalization of Theorem 4.14.

In [2], Alspach, Heinrich and Mohar demonstrated the Hamiltonicity of 1 block-intersection graphs of PBDs using a method which is very different from any of the

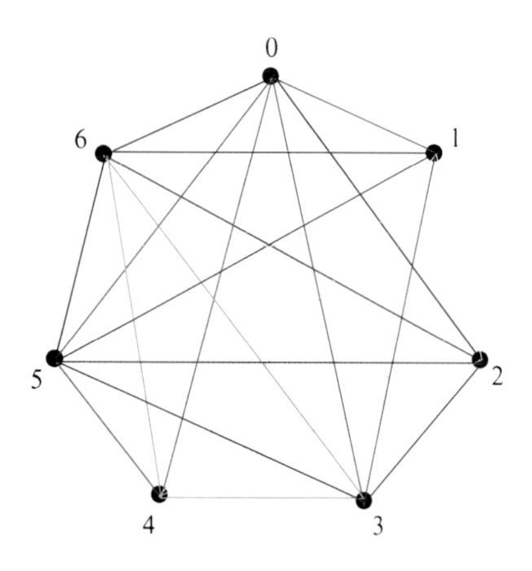

Fig. 4.3 G^V for the Fano plane

other Hamiltonicity results for block intersection graphs. It is necessary to introduce some additional terminology. Let $S = (V, \mathscr{B})$ be a block design. Recall that the block coloured pair adjacency graph of S is the multigraph with vertex set V such that, for each pair of distinct vertices v_1 and v_2, there is an edge between v_1 and v_2 of colour B_i if $\{v_1, v_2\} \subseteq B_i \in \mathscr{B}$. Denote this graph G^V. Note that when $\lambda = 1$, G^V is the complete graph on $|V|$ vertices. Recall that a trail is a walk in a multigraph in which no edge appears twice. A trail in G^V is called **block-dominating** if the set of vertices in the trail intersects every block of $\mathscr{B}$. For example, the trail 0, 1, 3 is block-dominating for the Fano plane (see Fig. 2.2 on page 17). Figure 4.3 illustrates G^V for the Fano plane.

Lemma 4.2 ([2]). *Let S be a block design. The block-intersection graph of S is Hamiltonian if and only if G^V has a block-dominating closed trail W with no two edges of W having the same colour.*

Note that in Fig. 4.3, the trail $0, 1, 3$ is block-dominating but has all edges the same colour; however, the trail $0, 1, 6, 3, 2$ is a block-dominating closed trail with no two edges having the same colour.

Proof. Let G_S denote the block-intersection graph of $S = (V, \mathscr{B})$.

($\Longrightarrow$) Suppose G_S has a Hamilton cycle, say $B_0, B_1, \ldots, B_{b-1}$. Let $a_i \in V$ be an element of $B_i \cap B_{i+1}$, where addition is performed modulo b. Consider the sequence $a_0, a_1, \ldots, a_{b-1}$ induced by the Hamilton cycle. For each contiguous subsequence $a_i, a_{i+1}, \ldots, a_{i+j}$ with $a_i = a_{i+1} = \cdots = a_{i+j}$, determine the maximal length of this subsequence and replace it by a single representative element. The closed trail obtained after applying this operation until it cannot be performed again is block-dominating (as the sequence contains an element from each block), and each of its edges has a different colour (simply use the rule of connecting elements a_i and a_{i+1} with edge colour B_i).

($\Longleftarrow$) Suppose G^V has a block-dominating closed trail $W = a_0, e_0, a_1, e_1, \ldots, a_\ell, e_\ell, a_0$, with no two edges having the same colour: Let B_i be the block

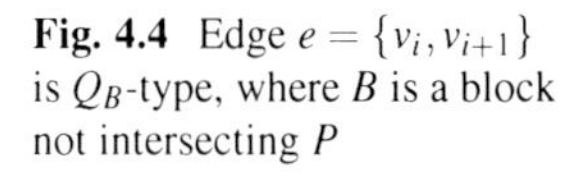

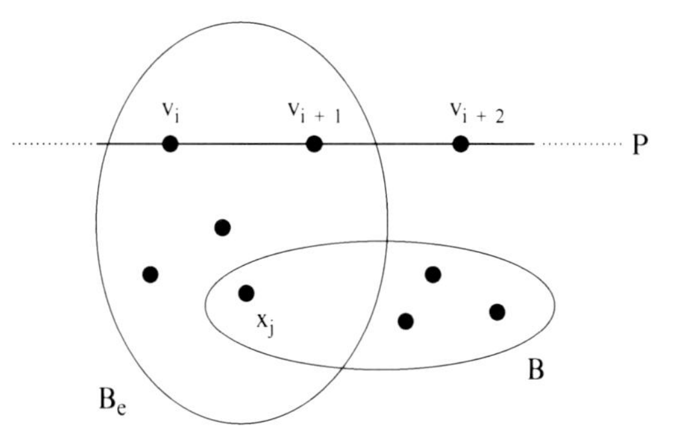

Fig. 4.4 Edge $e = \{v_i, v_{i+1}\}$ is Q_B-type, where B is a block not intersecting P

"containing" the edge e_i, that is, let $\{a_i, a_{i+1}\} \in B_i$. We construct a Hamilton cycle in G_S as follows. Let $C_1, C_2, \ldots, C_t \in \mathscr{B}$ be all the blocks containing a_0 whose colour does not appear in W. Then $P = C_1, C_2, \ldots, C_t, B_0$ is a trail in G_S. Let $C_{t+1}, C_{t+2}, \ldots, C_{t+r} \in \mathscr{B}$ be all the blocks containing a_1 which do not appear in the path P and whose colour does not appear in W. Since the edge e_0 joins a_0 to a_1, we know that both a_0 and a_1 appear in block B_0. Consequently, $P' = C_1, \ldots, C_t, B_0, C_{t+1}, \ldots, C_{t+r}, B_1$ is a trail in G_S. Continuing in this manner produces a Hamilton cycle in G_S due to the fact that together the elements of W touch every block of $\mathscr{B}$, so every block will eventually be included in the path. □

A little more terminology is required before proving the Hamiltonicity of certain PBDs. Let $S = (V, \mathscr{B})$ be a PBD$(v, K, 1)$. Let P be a trail in G^V with every edge a different colour. Suppose there are ℓ points in P and let n denote the number of distinct points in P. Clearly $n \le \ell$. Suppose that P is not block-dominating and let $B = \{x_0, x_1, \ldots, x_{k-1}\} \in \mathscr{B}$ be a block that does not intersect P. Let Q_B be the set of edges in P (which by definition have neither endpoint in B) representing blocks that intersect B, that is, $Q_B = \{e | e \in E(P), B_e \cap B \neq \emptyset\}$. The edge $e = \{v_i, v_{i+1}\}$ in Fig. 4.4 represents a Q_B-type edge.

Theorem 4.15 ([2]). *Every PBD$(v, K, 1)$, with* $\max(K) \le 2 \cdot \min(K)$, *admits a cyclic ordering in which every pair of consecutive blocks intersects in exactly one point.*

Proof. Assume $S = (V, \mathscr{B})$ is a non-trivial PBD$(v, K, 1)$ satisfying the block size restrictions. Let $M = \max(K)$ and $m = \min(K)$, then $M \le 2m$. We construct a block-dominating closed trail in G^V which, appealing to Lemma 4.2, implies the existence of a Hamilton cycle in the block-intersection graph of S.

Since S is non-trivial, there are three points, say $x, y, z \in V$, not all in the same block; hence, $P = x, y, z$ is a 3-cycle in G^V with all three edges of different colour. If P is block-dominating, we are done. Otherwise, we must extend P to a longer cycle with edges of different colours. We claim that we eventually obtain a block-dominating cycle.

Suppose P is a cycle in G^V, all of whose edges have different colours, and assume that $P = v_0, v_1, \ldots, v_{\ell-1}$. Let $B = \{x_0, x_1, \ldots, x_{k-1}\} \in \mathscr{B}$ be a block that does not intersect P, that is, $B \cap \{v_0, v_1, \ldots, v_{\ell-1}\} = \emptyset$. We give two methods for extending P to block-dominate B. Suppose there are two consecutive Q_B-type edges in P, $e_{i-1} = \{v_{i-1}, v_i\}$ and $e_i = \{v_i, v_{i+1}\}$. Without loss of generality, let $x_s \in B_{e_{i-1}}$ and $x_t \in B_{e_i}$, $s,t \in \{0,1,\ldots,k-1\}$, where $x_s \neq x_t$. We must have $x_s \neq x_t$ since if x_s were equal to x_t, the points v_i and x_s would appear in both $B_{e_{i-1}}$ and B_{e_i}, implying $B_{e_{i-1}} = B_{e_i}$ because $\lambda = 1$. This would imply that P has two edges of the same colour which is not true. P can be extended by replacing the path v_{i-1}, v_i, v_{i+1} with the path $v_{i-1}, x_s, x_t, v_{i+1}$. This replacement does not remove any edge colours from P and adds one new edge colour. Suppose there are no consecutive Q_B-type edges in P, but that the edge $e_i = \{v_i, v_{i+1}\}$ in P is a Q_B-type edge such that B_{e_i} intersects B in x_s, where $s \in \{0,1,\ldots,k-1\}$. If there is an edge $\{v_i, x_t\}$ or an edge $\{v_{i+1}, x_t\}$, $t \in \{0,1,\ldots,k-1\}$, where $t \neq s$, whose colour does not appear in P, then P can be extended to a longer cycle by replacing the path v_i, v_{i+1} with the path v_i, x_t, x_s, v_{i+1} or the path v_i, x_s, x_t, v_{i+1}. This insertion adds two new colours to P and removes none.

The proof rests on the claim that P can be extended using the methods explained above until a block-dominating cycle is obtained. Suppose that P cannot be extended and let B be a block that does not intersect P. We can assume that the Q_B-type edges in P form a matching, since if any two Q_B-type edges are adjacent in P, P can be extended. We can also assume that if $\{v_i, v_{i+1}\}$ is a Q_B-type edge in P, then the colours of the edges $\{v_i, x_t\}$ and $\{v_{i+1}, x_t\}$ already appear in P for every $x_t \in B$. Let q be the number of Q_B-type edges in P. It is necessary to deal with two cases: $q = 0$ and $q \neq 0$. The proof presented by Alspach, Heinrich and Mohar in [2] neglects the possibility that $q = 0$; however, it *is* possible that $q = 0$, and we deal with this case first. Since P does not dominate the block $B = \{x_0, x_1, \ldots, x_{k-1}\}$, no x_j, for $0 \leq j \leq k-1$, appears in P. Also, since there are no Q_B-type edges in P, none of the blocks induced by the edges of P intersect B. Each point in P appears together with each x_j, $0 \leq j \leq k-1$, in some block because $\lambda = 1$; therefore, we can extend P by replacing the edge $\{v_0, v_1\}$ with the path $\{v_0, x_0\}, \{x_0, x_1\}, \{x_1, v_1\}$. Each of these edges has a distinct colour, and none of these colours appear in P. This contradicts the assumption that P cannot be extended.

Now suppose $q \geq 1$. Let the sizes of the blocks corresponding to Q_B-type edges be denoted $k_0, \ldots, k_{q-1}$ and let the blocks represented by Q_B-type edges be denoted $B_{q,i}$, $i \in \{0,1,\ldots,q-1\}$. Since no two edges in P can be the same colour, no other edge with a colour the same as a Q_B-type edge can be added to P. Such edges exist between the vertices of P and elements of B. We want to know how many edges of colour $B_{q,i}$ go between $V(P)$ and B. Only one point of $B_{q,i}$ can be in B; otherwise, a pair of points would appear in two different blocks. How many points of $B_{q,i}$ can be in P? All of them except the point in which $B_{q,i}$ and B intersect. There are $|B_{q,i}| = k_i$ points in $B_{q,i}$, so there are at most $k_i - 1$ of these in P. The largest number of edges that cannot be considered for inclusion in P is

$$(k_0 - 1) + \cdots + (k_{q-1} - 1) = \sum_{i=0}^{q-1} k_i - q.$$

This means there are at least

$$k\ell - \left(\sum_{i=0}^{q-1} k_i - q\right) = k\ell + q - \sum_{i=0}^{q-1} k_i$$

edges between $V(P)$ and B whose colours do not appear in P, where $k\ell$ is the total number of edges between vertices of P and elements of B. Since there are q Q_B-type edges in P, the Q_B-type edges have $2q$ distinct endpoints (remember that no two Q_B-type edges are incident), and therefore all edges between $V(P)$ and B whose colours do not appear in P appear at the remaining $\ell - 2q$ vertices. The number of these edges is $k(\ell - 2q)$; hence,

$$(\ell - 2q)k \geq u \geq k\ell + q - \sum_{i=0}^{q-1} k_i,$$

where u is the number of edges between $V(P)$ and B whose colours do not already appear in P. Finally,

$$k\ell + q - \sum_{i=0}^{q-1} k_i \geq k\ell + q - qM,$$

so $(\ell - 2q)k - (k\ell + q - qM) \geq 0$ which yields $qM - q - 2qk \geq 0$. Therefore, we have $M - 1 - 2k \geq 0$, which implies $M \geq 2k + 1$. This is a contradiction to the condition that $M \leq 2m$, since the minimum value of k is m. We conclude that it is always possible to extend P when $q \geq 1$. □

Hare improved Theorem 4.15 by proving that the block-intersection graph of every PBD$(v, K, 1)$, with $\min(K) \geq 3$, is edge-pancyclic [18], which yields the following result. Note that some of the strength of the original result is lost in the translation to configuration ordering.

Theorem 4.16 ([18]). *Every PBD$(v, K, 1)$, with $\min(K) \geq 3$, admits a cyclic ordering in which every pair of consecutive blocks intersects in exactly one point.*

The proof of Theorem 4.16 requires dealing with many cases. Instead of presenting the entire proof, we outline the approach taken. Let $(V, \mathscr{B})$ be a PBD$(v, K, 1)$. Let $\ell = \min(K)$ and let $u = \max(K)$. Single out a specific block of the design $B^* \in \mathscr{B}$, $B^* = \{b_1, b_2, \ldots, b_k\}$ for some $k \in K$. For $i = 1, 2, \ldots, k$, let $\mathscr{B}_i = \{B \in \mathscr{B} : B \cap B^* = \{b_i\}\}$. Finally, let $\mathscr{B}^* = \mathscr{B}_1 \cup \mathscr{B}_2 \cup \ldots \cup \mathscr{B}_k$ and let G denote the 1 block-intersection graph for the PBD. The following lemma is used in the proof of Theorem 4.16.

Lemma 4.3 ([18]). *In the graph G, for $i, j \in \{1, 2, \ldots, k\}$, $i \neq j$ and $B \in \mathscr{B}_i$, the number of neighbours of B in $\mathscr{B}_j$ is $|B| - 1$ and the number of edges between $\mathscr{B}_i$ and $\mathscr{B}_j$ is $v - k$. Furthermore, $(v - k)/(u - 1) \leq |\mathscr{B}_i| \leq (v - k)/(\ell - 1)$.*

To prove Theorem 4.16, each case is divided into first constructing short cycles and then constructing long cycles that contain a particular edge. The following proposition is at the core of the proof of the theorem; it guarantees the existence of long paths which are used in the proof to construct long cycles.

Proposition 4.1. *Suppose* $(V,\mathscr{B})$ *is a* $PBD(v,K,1)$ *such that* $\ell > 3$ *and suppose* B^* *is any block of* $\mathscr{B}$ *such that* $|B^*| = u$. *Let* A^* *be any vertex of* $\mathscr{B}_2$, *let* $\mathscr{C} \subseteq \mathscr{B} \setminus (\mathscr{B}_2 \cup \mathscr{B}_3 \cup \ldots \cup \mathscr{B}_u)$ *and let* $H = G[\mathscr{B} \setminus (\mathscr{C} \cup \mathscr{B}_1 \cup \{\mathscr{B}^*\})]$. *For each* p, $1 \leq p \leq |V(H)| - 1$, *there is a path in* H *of length* p *that starts in* A^* *and ends in a vertex of* $\mathscr{B}^* \setminus \mathscr{B}_1$.

The interested reader is invited to see the details of the proof of Theorem 4.16 in [18].

It would be interesting to investigate the existence of A'_2-orderings for designs with $\lambda > 1$. Unfortunately, the work on pancyclicity for $\lambda > 1$ sheds no light on such orderings as block intersections may be between one and $\max(K)$.

The possible existence of A_3- and A_4-orderings is restricted to triple systems with $\lambda > 1$. In Sect. 5.2 we will show constructions for several infinite families of TTS(v)s admitting A_3-cyclic orderings. These results are in the language of 2-intersecting Gray codes; we restate Theorems 5.10 and 5.11 here in configuration ordering terms.

Corollary 4.5. *For each of the following orders there exists a TTS*(v) *that admits an* A_3*-cyclic ordering:*

- $v \equiv 1,4 \pmod{12}$, *where* $v \not\equiv 0 \pmod{5}$
- $v \equiv 3,7 \pmod{12}$, $v \geq 7$.

The existence of TTS(v)s with $v \equiv 0,6,9,10 \pmod{12}$ admitting A_3-orderings remains an open question. As M. Colbourn and Johnstone have shown that there exist TTS(v)s that do not admit A_3-orderings [12], we feel the best general result we can hope to prove is the existence of a TTS(v) of each admissible order that admits an A_3-cyclic ordering for its blocks. Corollary 4.5 is halfway to achieving this desired result.

The most natural generalization of the A_3 configuration to blocks of size k is the $(k+1,2)$-configuration—a configuration on two blocks in which the blocks share all but two points. Of course, other possible generalizations are the configurations in which two blocks overlap in m points, where $2 \leq m < k$, but such configurations are not necessarily of interest and the existence of such orderings is highly dependent on the relationship of m to k. Focusing on $(k+1,2)$-orderings, the existence of both Ucycles of rank three and 2-intersecting cyclic Gray codes for triple systems imply the existence of $(4,2)$-cyclic orderings for the blocks of triple systems. On the other hand, a $(k+1,2)$-cyclic ordering of the blocks of a BIBD(v,k,λ) translates into a Ucycle of rank k for the BIBD only in certain circumstances. For example, suppose a portion of a $(6,2)$-ordering for the blocks of a BIBD$(v,5,\lambda)$, with $\lambda \geq 4$, is $\{a,b,c,d,e\}$, $\{b,c,d,e,f\}$, $\{b,c,d,e,g\}$. To express this list in Ucycle form, we begin with the point a, then some ordering of the remaining points in the first block,

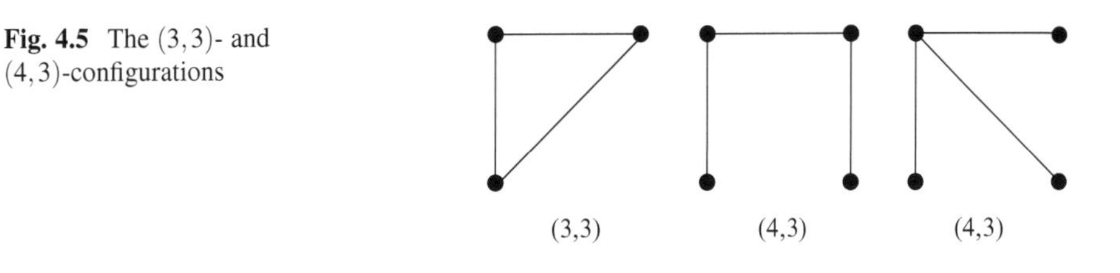

Fig. 4.5 The $(3,3)$- and $(4,3)$-configurations

followed by the point f. However, f is not in the third block, and so a Ucycle of rank five cannot be formed from this configuration ordering. We know of no existence results regarding $(k+1,2)$-orderings for BIBD(v,k,λ)s with $k>3$, except the following result—translated from minimal change combinatorial Gray codes—which applies to the family of designs consisting of all k-subsets of an n-set. These are simple designs with $|\mathscr{B}| = \binom{v}{k}$.

Theorem 4.17 ([3, 4, 6, 16, 31, 37]). *For $k \geq 2$, every simple BIBD$\left(v,k,\binom{v-2}{k-2}\right)$ admits a $(k+1,2)$-cyclic ordering.*

A_3-orderings (and A_2 orderings when $\lambda = 1$) are orderings with the fewest number of points in the union of any two consecutive blocks of the ordering. This idea can be extended to considering the number of points in the union of any d consecutive blocks. Orderings with a minimal number of induced points have applications to erasure-correcting codes for RAID disk arrays which we discuss in Sect. 6.1. All work in this area has been with respect to graphs, that is, blocks of size two. An ordering is **(d,f)-cluttered** if every set of d consecutive edges induces at most f points. This is not to be confused with configurations which are represented as $(points,\ edges)$. A cluttered ordering is **optimal** if f is the smallest admissible value given d.

Cohen, Colbourn and Fronček consider the existence of cluttered orderings for the complete graph in [9]. An edge ordering for K_v where three consecutive edges induce either three points (a triangle) or four points (either a path or star) and where at least $1/4(v^2-v-6)$ of the subgraphs formed are triangles is called a **ladder ordering of pairs**.

Theorem 4.18 ([8]). *A ladder ordering of pairs for K_v exists for all admissible v, except possibly when $v \in \{15,18,22\}$.*

Note that this is equivalent to the following generalized configuration ordering statement with the set of $\{(3,3),(4,3)\}$-configurations illustrated in Fig. 4.5.

Theorem 4.19 ([8]). *Every BIBD$(v,2,1)$ admits a $\{(3,3),(4,3)\}$-ordering, except possibly when $v \in \{15,18,22\}$.*

Regarding cluttered orderings in general, Cohen, Colbourn and Fronček have proved several interesting results:

Theorem 4.20 ([9]). *If a (d,f)-cluttered ordering for K_t exists, a (d',f')-cluttered ordering for K_{st+r} also exists, where $f' = min(sf+f, sf+r)$ and $d' \geq (s^2 + 2\lfloor \binom{s}{2}/(t-1)\rfloor)\cdot d$.*

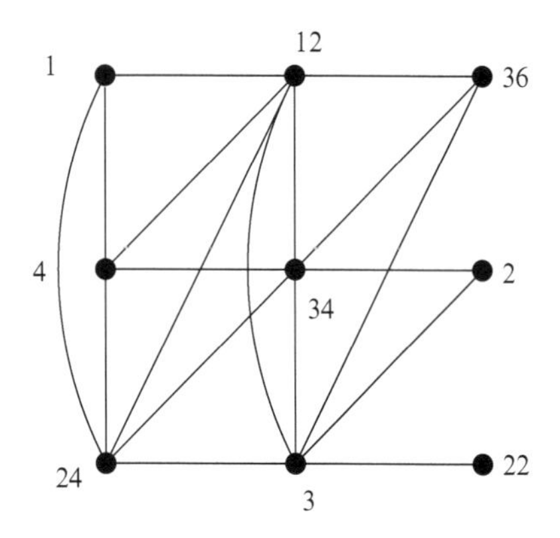

Fig. 4.6 A ρ-labelling of $G(3,2)$ in $\mathbb{Z}_{37}$

The major results in [9] are obtained by a special technique for producing specific (d,f)-cluttered orderings. Define a graph $G(h,w)$ on $h(w+1)$ vertices. The vertex set is $V = \{v_{i,j} : 0 \leq i < h, 0 \leq j \leq w\}$. The edge set is

$$E = \{\{v_{i,j}, v_{k,j}\} : 0 \leq i < k < h, 0 \leq j < w\} \cup \{\{v_{i,j}, v_{k,j+1}\} : 0 \leq k \leq i < h, 0 \leq j < w\}.$$

There are $w \cdot h^2$ edges. In fact, $G(h,w)$ has a natural edge partition into w copies of $G(1,h)$. A $\boldsymbol{\rho}$**-labelling** of a graph $G = (V,E)$ is a mapping $\rho : V \mapsto \mathbb{Z}_{2|E|+1}$ for which $\{\pm(\rho(r) - \rho(s)) : \{r,s\} \in E\}$ contains all non-zero numbers in $\mathbb{Z}_{2|E|+1}$. A ρ-labelling of a graph $G(h,w)$ is **wrapped** if it satisfies $\{\alpha + \rho(v_{i,0}) : 0 \leq i < h\} = \{\rho(v_{i,w}) : 0 \leq i < h\}$ for some α which is relatively prime to $2|E|+1$. Since the labelling hits every difference in $\mathbb{Z}_{2|E|+1}$, it can be developed in the group, and the developments include every possible edge in the complete graph on vertex set $\mathbb{Z}_{2|E|+1}$. The conditions on α guarantee that all the developed copies of the image of $G(h,w)$ can be ordered in a sequence that respects the configuration ordering. The graph $G(3,2)$ together with a ρ-labelling of the vertices with values from $\mathbb{Z}_{37}$ is shown in Fig. 4.6 (the figure appears in [9]). It is easy to check that the differences along edges include each difference from $\mathbb{Z}_{37}$ exactly once. Given these wrapped ρ-labellings, Cohen, Colbourn and Fronček go on to establish existence of certain cluttered orderings. For example,

Theorem 4.21 ([9]). *If $G(h,w)$ has a wrapped ρ-labelling, then there exists an $(h^2, 2h)$-cluttered ordering for K_{2wh^2+1}.*

Müller, Adachi and Jimbo have taken the work of Cohen et al. further by considering bipartite and tripartite graphs. They adapt the notion of a ρ-labelling to the bipartite graph as follows: A $\boldsymbol{\Delta}$**-labelling** of a bipartite graph with ℓ edges is a map from the vertices to a (perhaps smaller) bipartite graph with $\mathbb{Z}_\ell \times \{0,1\}$ where each edge induces a unique difference. In contrast to Cohen et al.'s requirement that the ρ-labelling be injective, Müller et al. permit non-injective Δ-labellings. They then combine these labellings with a base ordering to generate a full (d,f)-cluttered ordering through the action of a permutation group. Using three general constructions, they produce a series of orderings for complete bipartite graphs.

Theorem 4.22 ([28]). *For all $t \in \mathbb{N}$, there is a (d,f)-cluttered ordering for the complete bipartite graph $K_{3t,3t}$ with $d = 3s+r$, $f = 2(s+1)+r$, $s > 0$ and $r = 0,1,2$.*

Theorem 4.23 ([28]). *For all $t \in \mathbb{N}$, there is a (d,f)-cluttered ordering for the complete bipartite graph $K_{10t,10t}$ with $d = 10s+r$, $f = 4(s+1)+\min(r,4)$, $s > 0$ and $0 \le r \le 9$.*

Theorem 4.24 ([28]). *For all $h \in \mathbb{N}$, there is a (d,f)-cluttered ordering for the complete bipartite graph $K_{h(2h+1),h(2h+1)}$ with $d = h(2h+1)$ and $f = 4h$.*

From these results they can derive several optimal orderings.

Corollary 4.6 ([28]).

1. *For any $\ell \in \mathbb{N}$, where $3|\ell$, there exists an ordering of the edges of $K_{\ell,\ell}$ which is optimally $(3,4)$-, $(4,5)$-, $(5,6)$- and $(6,6)$-cluttered.*
2. *For any $\ell \in \mathbb{N}$, where $10|\ell$, there exists an optimal $(10,8)$-cluttered ordering for $K_{\ell,\ell}$.*

For example, the complete bipartite graph $K_{6,6}$, with vertex parts $\{0,1,2,3,4,5\}$ and $\{a,b,c,d,e,f\}$, can be optimally $(3,4)$-, $(4,5)$-, $(5,6)$- and $(6,6)$-cluttered ordered:

$$\{0,a\},\{0,b\},\{2,a\},\{2,b\},\{2,f\},\{5,b\},\{5,f\},\{5,a\},\{1,f\},\{1,a\},\{1,e\},$$
$$\{4,a\},\{4,e\},\{4,f\},\{0,e\},\{0,f\},\{0,d\},\{3,f\},\{3,d\},\{3,e\},\{5,d\},\{5,e\},$$
$$\{5,c\},\{2,e\},\{2,c\},\{2,d\},\{4,c\},\{4,d\},\{4,b\},\{1,d\},\{1,b\},\{1,c\},\{3,b\},$$
$$\{3,c\},\{3,a\},\{0,c\}.$$

In Sect. 6.1, we discuss the application of cluttered orderings to erasure-correcting codes for RAIDs. We will see that Müller et al. were motivated to generalize Cohen et al.'s work because of a specific code that is used in industry.

Adachi further extends the construction methods of [28] to tripartite graphs and uses them to construct cluttered orderings in $K_{9,9,9}$.

Theorem 4.25 ([1]). *There exists an optimal $(9,8)$-cluttered ordering for the complete tripartite graph $K_{9,9,9}$.*

The final two-block configuration to be considered is A_4 (see Fig. 1.1 on page 4). The only triple systems admitting A_4-orderings are those consisting of multiple (at least two) copies of a single block. These are the trivial designs: TS$(3,\lambda)$, $\lambda \ge 2$. Similarly, there are no interesting designs admitting A_4'-orderings.

We turn now to configurations involving three blocks. All three-block configurations in which each pair of blocks intersects in at most one point are shown in Fig. 1.2 (page 5). The same proof technique as used in proving Theorem 4.6 can be used to prove the existence of B_2-cyclic orderings.

Theorem 4.26 ([13]). *Every TS(v,λ), $v \ge 137$, admits a B_2-cyclic ordering.*

Proof. Let $S = (V, \mathscr{B})$ be a $\mathrm{TS}(v,\lambda)$. We employ the following result of Horák, Pike and Raines: The 1 block-intersection graph of any $\mathrm{TS}(v,\lambda)$, with $v \geq 12$ and arbitrary λ, is Hamiltonian [19] (Theorem 4.11). We have previously noted that a direct result of this theorem is that every $\mathrm{TS}(v,\lambda)$, $v \geq 12$, admits an A_2-cyclic ordering. More important to the current problem, the theorem implies that the blocks of any $\mathrm{TS}(v,\lambda)$, $v \geq 12$, can be decomposed into pairs, with the possible exception of one block, such that the two blocks making up a pair share exactly one element. Suppose $B_0, B_1, \ldots, B_{b-2}, B_{b-1}$ is a Hamilton cycle in the 1 block-intersection graph of S. We form pairs $\{B_0, B_1\}$, $\{B_2, B_3\}$, ... and so on. This sequence of pairs ends in either $\{B_{b-2}, B_{b-1}\}$ or in $\{B_{b-3}, B_{b-2}\}$ with B_{b-1} remaining.

Create a graph G with a vertex representing each of the pairs determined by a Hamilton cycle in the 1 block-intersection graph of S. If there is a block not paired, create a special vertex by amalgamating two vertices whose A_1 configurations intersect in a single point and whose blocks are disjoint from the leftover block. Include the leftover block in the amalgamated vertex (see Fig. 4.7). We must prove that such an arrangement of blocks is possible.

To create the amalgamated vertex we find four consecutive blocks in the Hamilton cycle that are disjoint from the triple leftover after decomposition. Suppose the block $B \in \mathscr{B}$ is not paired. Recall from the proof of Theorem 4.4 that the maximum number of blocks intersecting B is $3r - 2\lambda - 1$. Suppose we cannot find four consecutive blocks in the Hamilton cycle that do not intersect the leftover block. What is the maximum number of blocks that can be in such a Hamilton cycle? In decomposing the Hamilton cycle into A_2 configurations, we first remove the leftover block from the cycle, leaving a path where the first and last blocks intersect the leftover block in a single point. The worst possible arrangement in this path of the $3r - 2\lambda - 1$ blocks that intersect the leftover block is to have one appear every fourth block. Remembering that the path must start and end with intersecting blocks, we conclude that such a sequence contains $4(3r - 2\lambda - 2) + 1$ blocks. Therefore, in order to ensure that we are able to construct the amalgamated vertex, we require $4(3r - 2\lambda - 2) + 1 < b - 1$. In terms of v and λ, this requirement is $\lambda v^2 - 37\lambda v + 84\lambda + 36 > 0$. Thus, for all $v \geq 35$ and arbitrary λ, an amalgamated vertex can be formed when $|\mathscr{B}| \equiv 1 \pmod 2$. The amalgamated vertex has many possible forms; the three containing the largest number of vertices are given in Fig. 4.7.

To complete the construction of G, join two vertices of G by an edge if and only if the blocks represented by them are disjoint. We claim that a Hamilton cycle in G will provide a B_2-cyclic ordering for the blocks of S. Figure 4.8 displays a portion of the graph G.

To see that the ordering of vertices given by a Hamilton cycle in G is a B_2-cyclic ordering for the blocks they represent, consider the underlying blocks. Suppose the Hamilton cycle passes through v_0, v_{10}, v_4, v_6 (see Fig. 4.8). This segment of the cycle represents the block ordering $B_0, B_1, B_{10}, B_{11}, B_4, B_5, B_{14}, B_7, B_8, B_6, B_{12}$. The reader is invited to confirm that each set of three consecutive blocks in this list form a B_2 configuration. The order in which we list the blocks represented by a vertex is arbitrary, except when dealing with the amalgamated vertex. In general, the blocks

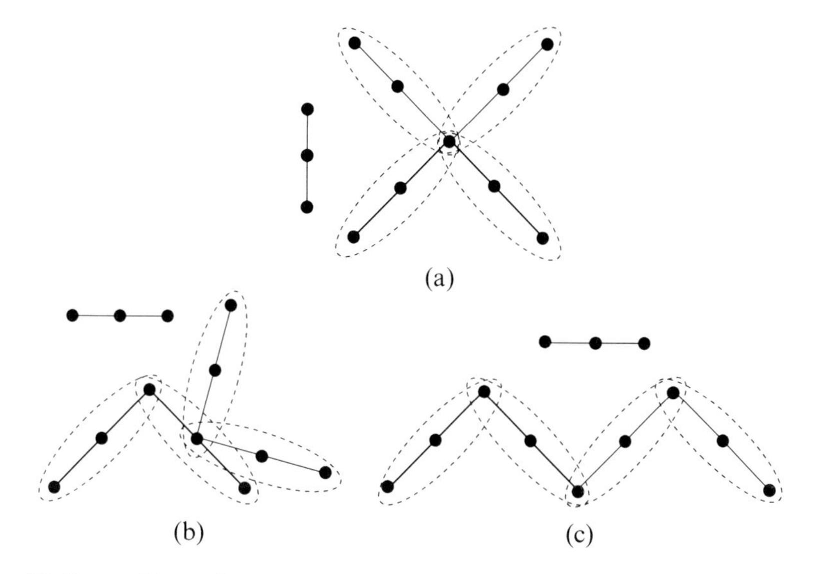

Fig. 4.7 Forms of the amalgamated vertex containing the largest possible number of points

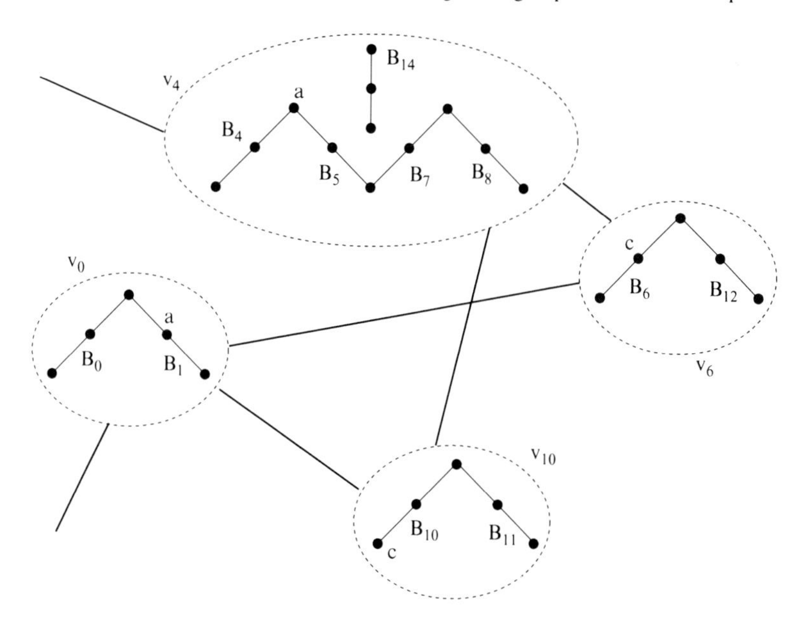

Fig. 4.8 A portion of G (for B_2-cyclic ordering)

of the amalgamated vertex will be listed as one of the A_2 configurations, followed by the disjoint block and then the remaining A_2 configuration. The amalgamated vertex of type (a) in Fig. 4.7 has four blocks that all intersect in the same point; therefore, any pairing of intersecting blocks may be selected.

To determine when G admits a Hamilton cycle, we appeal to Dirac's theorem (Theorem 2.2). In order to determine the smallest possible minimum degree of a vertex in G, we begin by considering the blocks represented by a given vertex and determining the maximum number of blocks of S that intersect these blocks. As usual, there are two cases to consider.

Case 1: $|\mathscr{B}| \equiv 0 \pmod 2$. When the number of blocks in the design is even, no amalgamated vertex appears in G. Using the same counting argument as seen in the proof of Theorem 4.6, we can determine that the maximum number of blocks that may intersect the blocks represented by a given vertex is $5r - 10\lambda + (10/3)\lambda - 2 = 5r - (20/3)\lambda - 2$. If each of these blocks appears as a member of a different pair of the decomposition (i.e. in a different vertex of G), then the given vertex is not adjacent to any of these vertices. The number of vertices in G is $v(v-1)\lambda/12$; therefore, the minimum degree of a vertex in G is $v(v-1)\lambda/12 - 1 - (5r - (20/3)\lambda - 2)$. Hence, $\delta(G) \geq v(v-1)\lambda/12 - (5/2)(v-1)\lambda + (20/3)\lambda + 1$.

To apply Dirac's theorem, we ask: when is $\delta(G)$ greater than or equal to $|V(G)|/2$? The answer is when $\lambda v^2 - 61\lambda v + 220\lambda + 24 \geq 0$. That is, the graph G induced by the triple system has a Hamilton cycle when this inequality holds. We conclude that every $\mathrm{TS}(v,\lambda)$, with $v \geq 58$ and an even number of blocks, admits a B_2-cyclic ordering.

Case 2: $|\mathscr{B}| \equiv 1 \pmod 2$. When the number of blocks in S is odd, an amalgamated vertex will appear in G. The amalgamated vertex has the largest potential number of blocks intersecting it due to the fact that it contains at most 12 distinct points (versus five for all other vertices). Using the same methodology as has been employed in previous proofs, we can determine that the maximum number of blocks that may intersect the blocks of the amalgamated vertex is $12r - 66\lambda + 22\lambda - 5 = 12r - 44\lambda - 5$. If each of these blocks appears as a member of a different pair of the decomposition (i.e. in a different vertex of G), then the amalgamated vertex is not adjacent to any of these vertices. The number of vertices in G is $(b-1)/2 - 1 = (v(v-1)\lambda - 18)/12$, since the amalgamated vertex contains two A_2 configurations. Therefore, the minimum degree of the amalgamated vertex is $(v(v-1)\lambda - 18)/12 - 1 - (12r - 44\lambda - 5)$. Hence, $\delta(G) \geq (v(v-1)\lambda - 18)/12 - 6(v-1)\lambda + 44\lambda + 4$.

Again we ask: when is $\delta(G)$ greater than or equal to $|V(G)|/2$? This occurs when $\lambda v^2 - 145\lambda v + 1200\lambda + 78 \geq 0$; thus, we conclude that every $\mathrm{TS}(v,\lambda)$, $v \geq 137$, admits a B_2-cyclic ordering. □

Horák and Rosa [21], and independently Dewar [13], have proved the following corollary.

Corollary 4.7. *Every* $STS(v)$, $v \geq 136$, *is* B_2*-cyclic orderable.*

Theorem 4.26 also leads to the following corollary. Again, as in Corollary 4.2, lower bounds can be improved for decomposition.

Corollary 4.8.

1. *Every $TS(v,\lambda)$, $v \geq 58$, can be decomposed into B_2 configurations.*
2. *Every $STS(v)$, $v \geq 57$, can be decomposed into B_2 configurations.*

A B'_2 configuration consists of two blocks intersecting in a single point, with a third block disjoint from both. Since every BIBD$(v,k,1)$ admits an A'_2-cyclic ordering, and hence a decomposition into A'_2 configurations, the same method as used in the proof of Theorem 4.26 can be used to determine the existence of B'_2-orderings for BIBD$(v,k,1)$s. As this proof is a generalization of one already presented, it is left to the diligent reader.

Theorem 4.27 ([13]). *Every BIBD$(v,k,1)$, with $v \geq 18k^2 - 6k + 1$, admits a B'_2-cyclic ordering.*

Corollary 4.9 ([13]). *Every BIBD$(v,k,1)$, with $v \geq 18k^2 - 6k + 1$, can be decomposed into B'_2 configurations.*

A B_3-ordering for the blocks of a triple system would require that one point appear in every block of the design. Indeed, consider four consecutive blocks in such an ordering and denote the blocks W, X, Y and Z. As each set of three consecutive blocks must intersect in a single point, we have $W \cap X \cap Y = \{x\}$. Since X and Y share the common point x (and no others), Z must also contain the point x (and no other points in $X \cup Y$) since X, Y and Z also form a B_3 configuration. This property implies that $r = b$, and hence, $v = 3$. However, the blocks of a TS$(3,\lambda)$ are copies of a single block; thus, they cannot form a B_3 configuration, let alone a B_3-ordering. The same argument applies for general k.

The form of configurations B_4 and B_5 is more promising than B_3; however, we find they are more difficult to deal with than B_1 and B_2. This difficulty stems from the fact that detailed information about how the first and third blocks in each consecutive set of three triples are related is required. Such information is not communicated by the 1 block-intersection graph of a triple system. We know of no results regarding the existence of B_4-orderings. A B_4-cyclic ordering for the blocks of a triple system would yield a rank two Ucycle for the blocks of that triple system; however, a rank two Ucycle is not necessarily (and, in fact, is unlikely to be) a B_4-ordering. In [7], Cohen and Colbourn prove that for each order $v \geq 15$, there exists an STS(v) that does not admit a B_5-ordering.

Theorem 4.28 ([7]). *For each admissible $v \geq 15$, there exists an $STS(v)$ which is not B_5-orderable.*

Proof. We first show that a Steiner space cannot have a B_5 ordering. A **Steiner space** is an STS such that every three elements which do not appear together in a triple are contained in a proper subsystem. A proper subsystem of an STS $(V,\mathscr{B})$ is a pair $(V',\mathscr{B}')$ with $V \subset V'$ and $\mathscr{B} \subset \mathscr{B}'$, $|V'| > 3$ and where $(V',\mathscr{B}')$ is itself an STS. Suppose $(V,\mathscr{B})$ is a Steiner space which does have a B_5 ordering. Consider two consecutive triples in this ordering, B_1, B_2. Suppose that $x \in B_1 \cap B_2$, $y \in B_1 \setminus B_2$ and $z \in B_2 \setminus B_1$. Then the three elements $\{x,y,z\}$ do not appear together in any triple and therefore must determine a proper subsystem. This subsystem must contain the two triples B_1 and B_2, and any triple preceding or following two consecutive triples of a subsystem must also lie in the subsystem because it will contain a pair of points from the union of the two consecutive triples. But this forces all triples of $\mathscr{B}$ to lie in the subsystem, which is a contradiction.

It is known that there exist Steiner spaces for all $v \equiv 1,3 \pmod 6$ except for $v \in \{19,21,25,33,37,43,51,67,69,145\}$ [11]. Thus, for each of these orders, there exists an STS which does not have a B_5 ordering. For the remaining ten cases, Cohen and Colbourn provide specific examples. □

Despite the statement of Theorem 4.28, Cohen and Colbourn expect that for every admissible $v \geq 15$, there *is* a B_5-orderable STS(v). In [7] they establish the existence of some small B_5-orderable STSs and hope that these orderings can be used in creating B_5-orderings for larger systems. In the next section, we relax the definition of configuration ordering by allowing each consecutive set of blocks of the specified number to be isomorphic to one of several different configurations. This will make the problem tractable and allow us to obtain orderings involving B_4 and B_5 configurations.

Because we have no methods for dealing with B_4-orderings nor B_5-orderings for triple systems, we have no methods for embarking on an investigation into the existence of configuration orderings for generalized versions of these configurations. In [20], Horák and Rosa noted that obtaining a result regarding the decomposition of BIBD$(v,k,1)$s into triangles—as generalized B_5 configurations are known—is likely to be difficult. Of course, finding a triangle ordering would imply the existence of such a decomposition, so such a result is likely to be even more difficult to obtain.

Finally, Cohen and Colbourn observe that no STS(v) can have a Pasch ordering because three blocks from a Pasch configuration uniquely determine the fourth block from the $(6,4)$-configuration; thus the same block would have to both precede and follow three consecutive blocks in such an ordering which is impossible for a design with index $\lambda = 1$ [7].

4.2 Generalized Configuration Ordering

As in standard configuration ordering, we attempt to proceed through collections of configurations using the naming convention of [11]. Jesso et al. have proved that the $\{1,2\}$ block-intersection graphs of various BIBDs are Hamilton-connected. This is equivalent to an $\{A_2', A_3'\}$-cyclic ordering.

Theorem 4.29 ([22,24]). *The $\{1,2\}$ block-intersection graph of any BIBD$(v,4,\lambda)$ with $v \geq 11$ and arbitrary λ is Hamilton-connected. Equivalently, every BIBD$(v,4,\lambda)$ with $v \geq 11$ admits an ordering in which every pair of consecutive blocks intersects in exactly one or two points.*

Theorem 4.30 ([22,24]). *The $\{1,2\}$ block-intersection graph of any BIBD$(v,5,\lambda)$ with $v \geq 57$ and arbitrary λ is Hamilton-connected. Equivalently, every BIBD$(v,5,\lambda)$ with $v \geq 57$ admits an ordering in which every pair of consecutive blocks intersects in exactly one or two points.*

Theorem 4.31 ([22,24]). *The $\{1,2\}$ block-intersection graph of any BIBD$(v,6,\lambda)$ with $v \geq 167$ and arbitrary λ is Hamilton-connected. Equivalently, every BIBD$(v,6,\lambda)$ with $v \geq 167$ admits an ordering in which every pair of consecutive blocks intersects in exactly one or two points.*

The proof methods are similar to Theorems 4.11, 4.12 and 4.13. Jesso et al. conjecture that corresponding results hold for all block sizes; we list these conjectures in Sect. 4.3.3.

We continue our survey of generalized configuration ordering results with another easy theorem due to basic graph theory. The result follows from the fact that K_v is Eulerian if and only if v is odd.

Theorem 4.32. *Every BIBD$(v,2,1)$, with v odd, admits a $\{(3,3),(4,3)_1\}$-cyclic ordering, where the $(4,3)_1$-configuration is the 4-path.*

It is only interesting to look at generalized configuration orderings where the set consists of configurations for which standard configuration orderings are not known to exist. For example, since every TS(v,λ), $v \geq 17$, admits an A_1-cyclic ordering (Theorem 4.4), if the set of configurations, $\mathscr{C}$, includes A_1, then these same triple systems admit $\mathscr{C}$-cyclic orderings. Since any set containing A_2 is also uninteresting and we have partial results implying the existence of A_3-cyclic orderings, we will not look at sets of configurations where each configuration consists of two blocks. For sets of configurations where each configuration consists of three blocks, we ignore sets that contain B_1 or B_2. The existence of a Hamilton cycle in the 1 block-intersection graph of every TS(v,λ), $v \geq 12$, is equivalent to the existence of a $\{B_3,B_4,B_5\}$-cyclic ordering for these triple systems. What can be said about orderings where the set consists of two configurations from the set $\{B_3,B_4,B_5\}$? The most logical pair to consider is $\{B_4,B_5\}$, as an ordering involving these configurations allows pairs of consecutive blocks to intersect in exactly one point and excludes sets of three consecutive blocks intersecting in a single point. A $\{B_4,B_5\}$-cyclic ordering for the blocks of a design is exactly a claw avoiding (or B_3-avoiding) Hamilton cycle in the 1 block-intersection graph of the design. A design admitting such an ordering is said to be **Eulerian**. The existence of such an ordering for STSs can be obtained by translating a result regarding the existence of rank two Ucycles (see Corollary 5.10 on page 162). Although a Ucycle of rank two is equivalent to a $\mathscr{C}$-cyclic ordering, where $\mathscr{C}$ is the set of configurations given in Fig. 5.10 (see page 167), when $\lambda = 1$, each pair of blocks intersects in at most one point and the set of configurations is reduced to $\{B_4,B_5\}$.

Corollary 4.10 ([13]). *Every cyclic STS(v), $v \neq 3$, admits a $\{B_4,B_5\}$-cyclic ordering.*

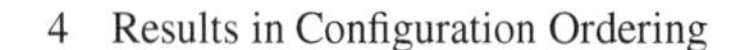

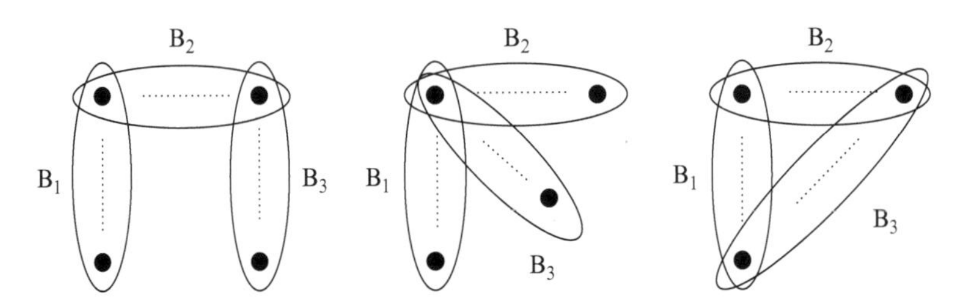

Fig. 4.9 The set of configurations in the translation of Hare's edge-pancyclicity result to configurations of three blocks

Hare's pancyclicity of the 1 block-intersection graph of certain PBDs can reasonably be translated into the generalized configuration language: every PBD$(v, K, 1)$, with $\min(K) \geq 3$, admits a $\{B_3', B_4', B_5'\}$-cyclic ordering (see Fig. 4.9). Of course, this result is better expressed as an A_2'-cyclic ordering since the original result is concerned only with the relationship of pairs of consecutive blocks, not triples of consecutive blocks. The results of Mamut, Pike and Raines [25] and Pike, Vandell and Walsh [29] can also be translated into the language of generalized configuration ordering; however, much clarity is lost as these results have consecutive blocks intersecting in a variable number of points.

The remaining results in generalized configurations come from translating results on the existence of Ucycles for designs. These results are dealt with in Chap. 5; therefore, the reader may wish to return to the results below after reading the proofs in Chap. 5.

Lemma 4.4 ([13]). *Every cyclic BIBD$(v,k,1)$, with $k \nmid v$, is $\{B_4', B_5'\}$-cyclic orderable.*

Proof. Let S be a BIBD$(v,k,1)$ with $k \nmid v$. The only problem base blocks that may appear in a cyclic BIBD$(v,k,1)$ are regular short orbit base blocks. Such base blocks have the form $\{0, v/k, 2v/k, \ldots, (k-1)v/k\}$; therefore, since k does not divide v, such base blocks do not appear in S. The design is generated entirely by full orbit base blocks. Applying the construction method described in Lemmas 5.8 and 5.9 to each of these base blocks yields a collection of partial $\{B_4', B_5'\}$-cyclic orderings representing the blocks developed. Since every pair of points appears in exactly one block, no double intersection of blocks will occur. Furthermore, the difference $1 \in \mathbb{Z}_v$ appears in a full orbit base block, and so there exists a length-v partial $\{B_4', B_5'\}$-cyclic ordering to which all other partial $\{B_4', B_5'\}$-cyclic orderings can be joined (in the manner described in the proof of Theorem 5.13) to form a $\{B_4', B_5'\}$-cyclic ordering for S. □

Another family of designs for which we understand the existence of $\{B_4', B_5'\}$-orderings are symmetric designs.

Lemma 4.5 ([13]). *Every symmetric cyclic BIBD$(v,k,1)$ is $\{B'_4, B'_5\}$-cyclic orderable.*

Proof. A symmetric cyclic design is generated by exactly one full orbit base block. Furthermore, when $\lambda = 1$, each difference in $\mathbb{Z}_v \setminus \{0\}$ appears exactly once in the base block; therefore, the base block is not a problem base block. The construction method described in the proof of Lemma 5.8 can be applied using the difference one. The resulting length-v partial $\{B'_4, B'_5\}$-cyclic ordering for the blocks developed from this base block is a $\{B'_4, B'_5\}$-cyclic ordering for the design. □

Lemma 4.6 ([13]). *The blocks of a symmetric BIBD(v,k,λ), $\lambda > 1$, cannot be $\{B'_4, B'_5\}$-ordered.*

Proof. In a symmetric BIBD(v,k,λ) (whether cyclic or non-cyclic) with $\lambda > 1$, every pair of blocks intersects in at least two points; hence, there is no ordering of the blocks that is a $\{B'_4, B'_5\}$-ordering. □

Finally, the existence of rank three Ucycles for some $TTS(v)$s (Theorem 5.10) can be translated into configuration ordering language where the configurations in question contain three blocks (see Fig. 4.10).

Corollary 4.11.

1. *For each* $v \equiv 1 \pmod{12}$, *with* $v \not\equiv 0 \pmod{5}$, *there exists a* $TTS(v)$ *that admits a* $\{B_6, B_7\}$*-ordering.*
2. *For each* $v \equiv 4 \pmod{12}$, *with* $v \not\equiv 0 \pmod{5}$, *there exists a* $TTS(v)$ *that admits a* $\{B_6, B_7\}$*-ordering.*
3. *For each* $v \equiv 7 \pmod{12}$, *there exists a* $TTS(v)$ *that admits a* $\{B_6, B_7\}$*-ordering.*

4.3 Graph Decomposition Designs

To this point, we have looked at configuration orderings for BIBDs and PBDs. Both of these types of design can be viewed as graph decompositions. Specifically, BIBD(v,k,λ)s are decompositions of K_v, with each edge repeated λ times, into copies of K_k. PBD(v,K,λ)s are similar, except the decomposition is into any K_k where $k \in K$. In this section, we discuss configuration orderings for other graph

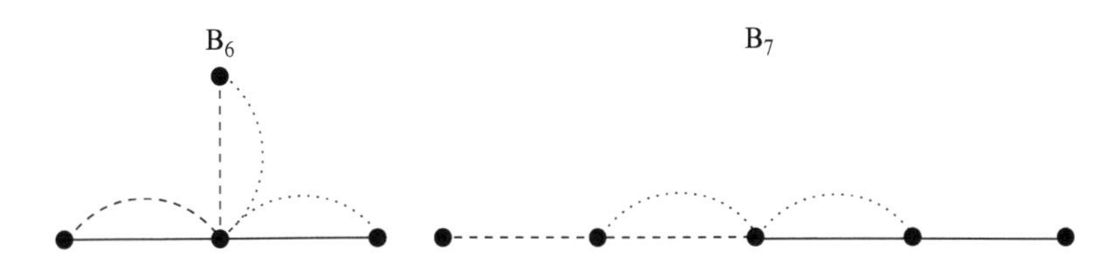

Fig. 4.10 The configurations B_6 and B_7

decompositions. Orderings are known for decompositions into matchings, decompositions into partial matchings and cycles, and near Hamilton decompositions.

4.3.1 Decompositions into Matchings

The **uniform 1-factorization problem** is to decompose a complete graph, K_{2n}, into 1-factors (matchings) such that the union of any two of them is isomorphic to a given graph (this consistency is what uniform refers to). If the given graph is a Hamilton cycle, the 1-factorization is said to be **perfect**. Perfect 1-factorizations are known whenever n or $2n-1$ is prime, and for a finite list of other values [15]. However, since the uniform 1-factorization problem is very difficult, Dinitz, Dukes and Stinson have relaxed the definition to ask for an ordering of the 1-factors such that the union of any two *consecutive* 1-factors is uniform.

Definition 4.1 (sequentially uniform 1-factorization). A 1-factorization of K_{2n} is sequentially uniform if there exists a sequence of the 1-factors, $\Upsilon = F_0, \ldots, F_{2n-2}$, where $F_i \cup F_{i+1} \cong F_j \cup F_{j+1}$, for all $0 \leq i, j \leq 2n-2$, where index addition is performed modulo $2n-1$.

A 1-factorization has (multiset) **type** $T = (k_1, \ldots, k_r)$ if $F_i \cup F_{i+1}$ is isomorphic to the disjoint union of cycles of lengths $k_1, \ldots, k_r$, where $\sum_i k_i = 2n$. When the type $T = (2n)$, the 1-factorization is said to be **sequentially perfect**.

A **starter** in $\mathbb{Z}_{2n-1}$ is a set of $n-1$ pairs $S = \{\{x_1, y_1\}, \ldots, \{x_{n-1}, y_{n-1}\}\}$ such that every non-zero element of $\mathbb{Z}_{2n-1}$ appears exactly once as some x_i or y_i and also exactly once as some difference $x_j - y_j$ or $y_j - x_j$ [14]. Let S^* denote $S \cup \{\{0, \infty\}\}$ and let $S^* + x$ denote the addition of x to each element in each pair of S^*, where $x + \infty = \infty + x = \infty$. Then $\{S^* + x \mid x \in \mathbb{Z}_{2n-1}\}$ is a 1-factorization of K_{2n} with vertex set $\mathbb{Z}_{2n-1} \cup \{\infty\}$. Dinitz et al. observe the following.

Lemma 4.7 ([15]). *Let S be a starter in $\mathbb{Z}_{2n-1}$ with $n \geq 1$. The ordered 1-factorization of K_{2n} generated by S: $S^*, S^*+1, S^*+2, \ldots, S^*+(2n-2)$ is sequentially uniform.*

Perhaps the most famous starter is the **patterned starter** $P = \{\{x, -x\} \mid x \in \mathbb{Z}_{2n-1}\}$. The 1-factor represented by the patterned started for $n = 7$ is shown in Fig. 4.11.

In [15], Dinitz et al. determine the type of the 1-factorization induced by the patterned starter for all $n \in \mathbb{N}$.

Lemma 4.8 ([15]). *Let P be the patterned starter in $\mathbb{Z}_{2n-1}$. Let $k \in \mathbb{Z}_{2n-1} \setminus \{0\}$ with* $\gcd(2n-1, k) = d$. *Then $P^* \cup (P^* + k)$ consists of one cycle of length $1 + (2n-1)/d$ and $(d-1)/2$ cycles of length $2(2n-1)/d$.*

Taking k relatively prime to $2n-1$ and combining Lemmas 4.7 and 4.8 yields the following theorem.

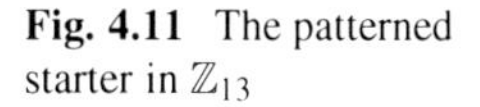
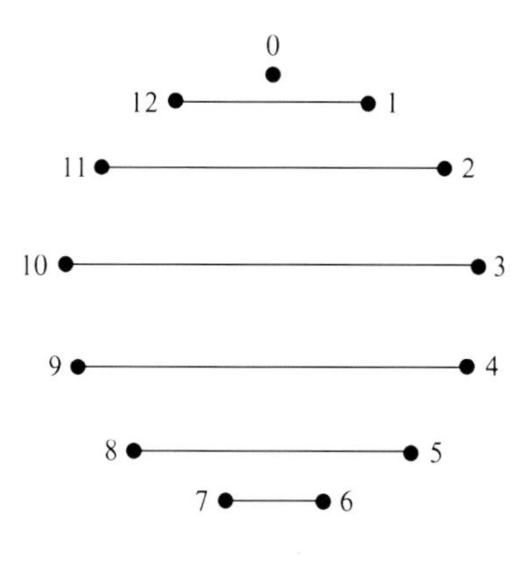

Fig. 4.11 The patterned starter in $\mathbb{Z}_{13}$

Theorem 4.33 ([15]). *For every $n \geq 1$, there exists a sequentially perfect 1-factorization of K_{2n}.*

Dinitz et al. go on to show that every possible type is realizable (sometimes trivially so) for K_{2n}, for $4 \leq 2n \leq 24$. These are not always constructed from starters. They also investigate starters in non-cyclic groups.

Theorem 4.34 ([15]). *Suppose $q = p^d$ is an odd prime power (with p prime and $d > 1$) such that $q = 2^{\ell}t + 1$ and $t > 1$ is odd. If p is not a Mersenne prime, or p is a Mersenne prime but $d > 2$, then there exists a sequentially uniform 1-factorization of K_{q+1}.*

Dinitz et al. use a product construction to show the following.

Theorem 4.35 ([15]). *For any odd integer $t \geq 1$, there is a sequentially uniform 1-factorization of $K_{2^{\ell}t}$ of type $(4,4,\ldots,4)$ for all integers $\ell \geq 2 + \lceil \log_2 t \rceil$.*

Venkaiah and Ramanjaneyulu have constructed sequentially perfect 1-factorizations of K_{2^n}.

Theorem 4.36 ([35]). *For every $\ell \geq 1$, there exists a sequentially perfect 1-factorization of $K_{2^{\ell}}$.*

4.3.2 Decompositions into Partial Matchings and Cycles

In sports tournament scheduling, the set of c games played in a particular time slot involving two competitors from a set of v competitors is often represented by a partial matching with c edges in K_v. A decomposition of K_v into partial matchings having c edges then represents a schedule for a full round-robin tournament. The necessary conditions for such a decomposition to exist are

$$c \mid \binom{v}{2} \tag{4.12}$$

$$1 \le c \le \left\lfloor \frac{v}{2} \right\rfloor \tag{4.13}$$

$$v - 1 + (v \pmod 2) \le \frac{\binom{v}{2}}{c}. \tag{4.14}$$

In this section, we state only the mathematical results on ordering these designs which can be expressed in configuration language. More discussion, proofs and the scheduling application can be found in Sect. 6.2.

An **optimum interval** is defined to be

$$s(v,c) = \begin{cases} \max(0, (\frac{v}{2c} - 2)) & \text{if } v \equiv 0 \pmod{2c} \\ \left\lfloor \frac{v}{2c} \right\rfloor - 1 & \text{otherwise.} \end{cases}$$

As the name implies, if an ordering of partial matchings can be given such that no edge appears in two matchings closer than the optimum interval apart, then games between each pair of players are as widely spaced as possible. It is impossible for any ordering of the c-matchings from a decomposition of K_v to have s consecutive matchings disjoint for $s > s(v,c) + 1$ (this is proved in Sect. 6.2). However, Rodney showed that the optimum interval can always be achieved.

Theorem 4.37 ([30]). *Whenever the necessary conditions are satisfied, there exists a decomposition of K_v into c-matchings and an ordering of the partial matchings so that any consecutive set of $s(v,c) + 1$ of them are vertex (and edge) disjoint.*

When $c = 2$, Rodney was also able to 2-edge colour each matching so that every point is incident with the same number ($\alpha = (v-1)/c$) of edges of each colour.

Theorem 4.38 ([30]). *Whenever the necessary conditions are satisfied and n is odd, there exists a decomposition of K_v into 2-matchings and an ordering of the partial matchings so that any consecutive $s(v,2) + 1$ of them are vertex (and edge) disjoint. Furthermore, the edges can be 2-coloured so that each partial matching has one edge of each colour and every vertex is incident to exactly $(v-1)/2$ edges of each colour.*

When $c > 2$, Rodney was not able to maintain an ordering in which every set of $s(v,c) + 1$ consecutive c-matchings were disjoint while simultaneously c-edge colouring; however, he could construct orderings in which every set of $s(v,c)$ consecutive c-matchings were disjoint.

Theorem 4.39 ([30]). *Whenever the necessary conditions are satisfied and $v \equiv 1 \pmod c$, there exists a decomposition of K_v into c-matchings and an ordering of the partial matchings so that any consecutive $s(v,c)$ of them are vertex (and edge) disjoint. Furthermore, the edges can be c-coloured so that each partial matching has one edge of each colour and every vertex is incident to exactly $(v-1)/c$ edges of each colour.*

De Werra has also investigated ordering partial c-matchings of K_v.

Theorem 4.40 ([36]). *If $c = (v-1)/2$, there is a decomposition of K_v into c-matchings, an ordering of every edge and an ordering of the partial matchings with the property that for any two consecutive appearances of a given vertex in an edge of a c-matching, the vertex appears in different positions in the two edges.*

Note that the two occurrences of a given vertex referred to in Theorem 4.40 may be in consecutive c-matchings in the ordering, or they may be separated by one c-matching in which the vertex is not incident with any edges. For $c = v/2$ we can do nearly as well.

Theorem 4.41 ([36]). *If v is even, there is a decomposition of K_v into c-matchings, an ordering of every edge and an ordering of the partial matchings with the property that, with $v-1$ exceptions, for any two consecutive appearances of a given vertex in an edge of a c-matching, the vertex appears in different positions in the two edges.*

It is impossible to avoid the $v-1$ exceptions referred to in Theorem 4.41. These orderings are used in balanced home and away tournament scheduling and are discussed more fully in Sect. 6.2.4.

4.3.3 Maximally Separated Twofold Near Hamilton Decompositions

In the context of scheduling matches for a small tennis club, Stevens proved the existence of a design with an edge-disjoint ordering property, which he calls edge-separated. The application is discussed further in Sect. 6.2.3. In a $\lambda = 2$ decomposition of K_v into $(v-1)$-cycles, if it is possible to cyclically order the v decompositions such that any s consecutive $(v-1)$-cycles are pairwise edge disjoint, then $s < \lfloor v/2 \rfloor$. Stevens proves that this optimum can always be obtained.

Theorem 4.42 ([33]). *For every $v \geq 4$, there exists a $\lfloor (v-2)/2 \rfloor$-edge-separated $\lambda = 2$ decomposition of K_v into cycles of length $v-1$. In fact, for v odd and $v \geq 9$, each cycle will have precisely two edges at separation $(v-3)/2$; all others will be maximally separated by $(v-1)/2$.*

Proof. The proof is by construction, and we give one of the cases: $v \equiv 1 \pmod 8$, $v \geq 75$. We present a base cycle that uses every difference in $\mathbb{Z}_v$ exactly twice and will be developed cyclically in $\mathbb{Z}_v$ to produce all the cycles in the design. The first edge of length two appears between vertices $(v-5)/4$ and $(v+3)/4$. The second edge of length two appears $(v-3)/2$ positions further in $\mathbb{Z}_v$. This is the only edge to appear $(v-3)/2$ positions further, all other edges appear precisely $(v-1)/2$ positions further in $\mathbb{Z}_v$. To save space, for the remainder of the edges, we will only note the length (difference), d, and the starting vertex, x, of the first instance edge. Its other endpoint will be $x+d$, and the other instance of this edge length d will occur precisely $(v-1)/2$ positions further in $\mathbb{Z}_v$.

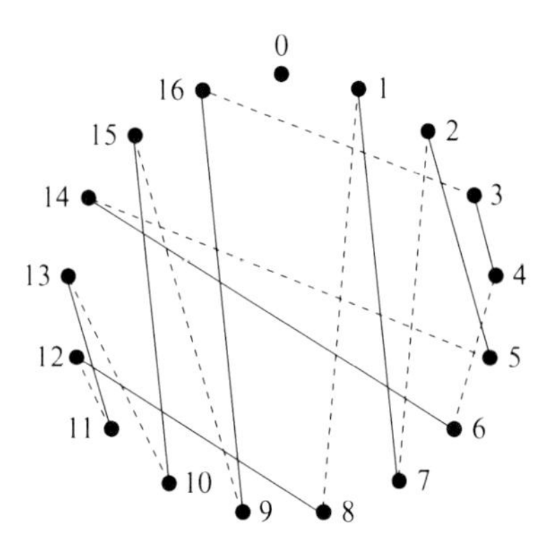

Fig. 4.12 A base cycle for a maximally separated twofold near Hamilton decomposition of K_{17}

The edge of length $(v-1)/2$ starts at vertex $(v+3)/4$. The edges of lengths $(v-1)/4 \leq \ell \leq (v-3)/2$ start at vertex $\lfloor (v-2\ell+1)/4 \rfloor$. The edge of length $(v-5)/4$ starts at vertex $(v-1)/2$, and the edge of length three starts at $(3v-27)/8$. For the remaining edges, define

$$i_{\max} = \begin{cases} (v-41)/24 & \text{if } v \equiv 17 \pmod{24} \\ (v-49)/24 & \text{if } v \equiv 1 \pmod{24} \\ (v-81)/24 & \text{if } v \equiv 9 \pmod{24}. \end{cases}$$

The edges of lengths in the range

$$(v-29)/4 - 6i \leq \ell \leq (v-9)/4 - 6i,$$

for $0 \leq i < i_{\max}$, are as follows: The edges of lengths $(v-9)/4-6i$ and $(v-9)/4-5-6i$ start at vertex $(v+7)/8+3i$; edges of lengths $(v-9)/4-1-6i$ and $(v-9)/4-2-6i$ start at vertex $(v+7)/8+1+3i$; and edges of lengths $(v-9)/4-3-6i$ and $(v-9)/4-4-6i$ start at vertex $(v+7)/8+2+3i$. An example for $v=17$ is given in Fig. 4.12. □

Conjectures

Jesso, Pike and Shalaby conjecture that results like Theorems 4.11–4.13 and Theorems 4.29–4.31 hold for any block size, k:

Conjecture 4.1 ([24]). For each $k \geq 3$, there exists a constant c_k such that for all $v \geq c_k$ and $\lambda \geq 1$, the 1 block-intersection graph of any BIBD(v,k,λ) is Hamiltonian.

Conjecture 4.2 ([24]). For each $k \geq 3$, there exists a constant c_k such that for all $v \geq c_k$ and $\lambda \geq 1$, the $\{1,2\}$-block-intersection graph of any BIBD(v,k,λ) is Hamiltonian.

Exercises and Problems

Exercise 4.1. Complete the details of the proof of Theorem 4.1.

Exercise 4.2. Complete the details of the proof of Corollary 4.2.

Exercise 4.3. Complete the details of the proof of Theorem 4.9.

Exercise 4.4. Prove Lemma 4.3.

Exercise 4.5. Prove Theorem 4.27.

There remain many open questions regarding the existence of configuration orderings; here are a few to consider:

Problem 4.1. Do $\mathrm{TS}(v,\lambda)$s admit B_4-orderings? Do $\mathrm{TS}(v,\lambda)$s admit B_5-orderings?

Problem 4.2. Can the known results in standard configuration ordering for triple systems be extended to $\mathrm{BIBD}(v,k,\lambda)$s?

Problem 4.3. Determine a lower bound on the number of edges at separation $(n-3)/2$ in *any* maximally separated twofold near Hamilton decomposition (as opposed to the cyclic ones which were constructed in Theorem 4.42).

References

1. Adachi, T.: Optimal ordering for the complete tripartite graph $K_{9,9,9}$. In: Nonlinear Analysis and Convex Analysis, edited by W. Takahashi and T. Tanaka, pp. 1–10. Yokohama Publ., Yokohama (2007)
2. Alspach, B., Heinrich, K., Mohar, B.: A note on Hamilton cycles in block-intersection graphs. In: Finite Geometries and Combinatorial Designs (Lincoln, NE, 1987), Contemporary Mathematics, vol. 111, pp. 1–4. American Mathematical Society, Providence, RI (1990)
3. Bitner, J.R., Ehrlich, G., Reingold, E.M.: Efficient generation of the binary reflected Gray code and its applications. Comm. ACM **19**(9), 517–521 (1976)
4. Buck, M., Wiedemann, D.: Gray codes with restricted density. Discrete Math. **48**, 163–171 (1984)
5. Case, G.A., Pike, D.A.: Pancyclic PBD block-intersection graphs. Discrete Math. **308**, 896–900 (2008)
6. Chase, P.J.: Combination generation and Graylex ordering. Congr. Numer. **69**, 215–242 (1989)
7. Cohen, M.B., Colbourn, C.J.: Optimal and pessimal orderings of Steiner triple systems in disk arrays. Theoret. Comput. Sci. **297**, 103–117 (2003)
8. Cohen, M.B., Colbourn, C.J.: Ladder orderings of pairs and RAID performance. Disc. Appl. Math. **138**(1), 35–46 (2004)
9. Cohen, M.B., Colbourn, C.J., Frončcek, D.: Cluttered orderings for the complete graph. In: Lecture Notes in Computer Science, vol. 2108, pp. 420–431 (2001)
10. Colbourn, C.J., Dinitz, J.H. (eds.): Handbook of Combinatorial Designs, second edn. Chapman & Hall/CRC, Boca Raton, FL (2007)
11. Colbourn, C.J., Rosa, A.: Triple Systems. Oxford Mathematical Monographs. The Clarendon Press Oxford University Press, New York (1999)

12. Colbourn, M.J., Johnstone, J.K.: Twofold triple systems with a minimal change property. Ars Combin. **18**, 151–160 (1984)
13. Dewar, M.: Gray codes, universal cycles and configuration orderings for block designs. Ph.D. thesis, Carleton University, Ottawa, ON (2007)
14. Dinitz, J.H.: Starters, chap. VI.55, pp. 622–628. In: Colbourn and Dinitz [10] (2007)
15. Dinitz, J.H., Dukes, P., Stinson, D.R.: Sequentially perfect and uniform one-factorizations of the complete graph. Electron. J. Combin. **12**, Research paper 1, (electronic, 12 pp.) (2005)
16. Eades, P., Hickey, M., Read, R.C.: Some Hamilton paths and a minimal change algorithm. J. Assoc. Comput. Mach. **31**(1), 19–29 (1984)
17. Harary, F., Robinson, R.W., Wormald, N.C.: Isomorphic factorisations I: complete graphs. Trans. Amer. Math. Soc. **242**, 243–260 (1978)
18. Hare, D.R.: Cycles in the block-intersection graph of pairwise balanced designs. Discrete Math. **137**, 211–221 (1995)
19. Horák, P., Pike, D.A., Raines, M.E.: Hamilton cycles in block-intersection graphs of triple systems. J. Combin. Des. **7**(4), 243–246 (1999)
20. Horák, P., Rosa, A.: Decomposing Steiner triple systems into small configurations. Ars Combin. **26**, 91–105 (1988)
21. Horák, P., Rosa, A.: Private communication (2005)
22. Jesso, A.: The Hamiltonicity of block-intersection graphs. Master's thesis, Memorial University of Newfoundland, St. John's, NF (2010)
23. Jesso, A.T.: Private communication (2011)
24. Jesso, Andrew T.(3-NF); Pike, David A.(3-NF); Shalaby, Nabil(3-NF) Hamilton cycles in restricted block-intersection graphs. (English summary) Des. Codes Cryptogr. **61**(3), 345–353 (2011)
25. Mamut, A., Pike, D.A., Raines, M.E.: Pancyclic BIBD block-intersection graphs. Discrete Math. **284**, 205–208 (2004)
26. Momihara, K., Jimbo, M.: Some constructions for block sequences of Steiner quadruple systems with error-correcting consecutive unions. J. Combin. Des. **16**(2), 152–163 (2008)
27. Momihara, K., Jimbo, M.: On a cyclic sequence of a packing by triples with error-correcting consecutive unions. Util. Math. **78**, 93–105 (2009)
28. Müller, M., Adachi, T., Jimbo, M.: Cluttered orderings for the complete bipartite graph. Discrete Appl. Math. **152**(1–3), 213–228 (2005)
29. Pike, D.A., Vandell, R.C., Walsh, M.: Hamiltonicity and restricted block-intersection graphs of t-designs. Discrete Math. **309**, 6312–6315 (2009)
30. Rodney, P.: Balance in tournament designs. Ph.D. thesis, University of Toronto, Toronto, ON (1993)
31. Ruskey, F.: Adjacent interchange generation of combinations. J. Algorithms **9**(2), 162–180 (1988)
32. Simmons, G.J., Davis, J.A.: Pair designs. Comm. Statist. **4**, 255–272 (1975)
33. Stevens, B.: Maximally pair separated round robin tournaments: ordering the blocks of a design. Bull. Inst. Combin. Appl. **52**, 21–32 (2008)
34. Varma, B.N.: On pancyclic line graphs. Congr. Numer. **54**, 203–208 (1986)
35. Venkaiah, V.C., Ramanjaneyulu, K.: Sequentially perfect 1-factorization and cycle structure of patterned factorization of K_{2^n} (2011). Presented at CanaDAM 2011 conference, Victoria, BC.
36. de Werra, D.: Some models of graphs for scheduling sports competitions. Discrete Appl. Math. **21**(1), 47–65 (1988)
37. Wilf, H.S.: Combinatorial algorithms: an update. CBMS-NSF Regional Conference Series in Applied Mathematics, 55. Society for Industrial and Applied Mathematics (SIAM), Philadelphia, PA (1989)

Chapter 5
Results in Gray Codes and Universal Cycles for Designs

In this chapter we consider Gray codes and universal cycles (Ucycles) for designs. The chapter is broken into three sections by type of design: minimal change designs, twofold triple systems, and cyclic BIBDs. In addition to proving the existence of Gray codes and Ucycles for certain designs, we discuss how these results relate to configuration orderings.

Prior to the formal introduction of definitions for Gray codes and Ucycles for designs (in the thesis of Dewar [9]), the terminology used to describe minimal change orderings for the blocks of a design was variable and the results scattered. Most was known about the existence of Gray codes for designs. Some Gray code results existed under the names single-change covering designs and serial treatment designs. These names are indicative of the fact that an ordering of the blocks is built directly into the definition of these designs. Single-change covering designs will be discussed in Sect. 5.1, while serial treatment designs will be discussed in terms of their application in Sect. 6.5.2. In other cases, the terminology does not necessarily indicate the ordering idea. For example, the Hamiltonicity of the 1 block-intersection graph for certain designs is equivalent to the existence of 1-intersecting Gray codes for these designs. These results were discussed in Chap. 4 in the context of configuration orderings.

The only Ucycle ordering result for non-trivial designs, prior to the work of Dewar, came from sequential covering designs—the Ucycle variant of single-change covering designs. These will be discussed in Sect. 5.1. We should note the existence of Ucycles for some k-subsets of n-sets are Ucycles for (not particularly interesting) block designs. These were discussed in Sect. 2.3.4.2. In Sects. 5.2 and 5.3 all results on the existence of Gray codes and Ucycles for designs are presented in this terminology. Of course, any rank k Ucycle yields the weaker $(k-1)$-intersecting Gray code, while the reverse is certainly not the case. Conversely, a rank t Ucycle, where t is the strength of the design in question, may induce a $(t-1)$-intersecting Gray code; however, much depends on the design in question, particularly on the relationship of λ to t. We will point out these connections

M. Dewar and B. Stevens, *Ordering Block Designs: Gray Codes, Universal Cycles and Configuration Orderings*, CMS Books in Mathematics,
DOI 10.1007/978-1-4614-4325-4_5,

throughout this chapter, in addition to translating results into configuration ordering terms where applicable; however, it will profit the reader to keep the idea of interlinked definitions foremost in their mind.

5.1 Minimal Change Designs

In this section, we look at designs whose very definition includes an order on the blocks. Recall that we are interested in block ordering, not ordering within blocks (see Chap. 1 for references to this other type of design). Single-change covering designs (SCCDs) and their variants are the only minimal change designs we know of. These designs are defined in Sect. 3.2.1, and in this section, we present select results in this area. Most work on these designs (namely, determining their existence) was done by a small group of authors throughout the 1990s.

Single-change covering designs were first introduced by Wallis, Yucas and Zhang in [24]; however, they had been investigated previously in largely unpublished work. An entertaining history of the work on single-change designs is given in [19].

Wallis, Yucas and Zhang present various existence and lower bound results. Recall that a SCCD(v,k) is economical if it has

$$\left\lceil \frac{\binom{v}{2}-\binom{k-1}{2}}{k-1} \right\rceil$$

blocks.

Lemma 5.1 ([24]). *There is an economical SCCD(v,k) whenever $k=v$, $k=v-1$ or v is odd and $k=2$. In fact, the number of blocks in these designs are 1, 3 and $v(v-1)/2$, respectively.*

Proof. The case $k=v$ is trivial and a SCCD$(v,v-1)$ is

$$\{1,2,3,\ldots,v-2,v-1\},\{1,2,3,\ldots,v-2,v\},\{2,3,4,\ldots,v-1,v\}.$$

When $k=2$, think of the points as vertices and the blocks as edges of K_v. A SCCD is a walk through the graph that covers every edge. When v is odd, K_v is Eulerian and any Euler trail forms an economical SCCD$(v,2)$. □

When v is even, one can take the sequence of blocks $\{1,2\},\{1,3\},\ldots,\{1,v\}$ followed by the edges of an Euler trail in K_{v-1} where vertices are labelled $2,3,\ldots,v$. The number of blocks is at most $v(v-1)/2$ which is clearly best possible.

Many small examples are presented in [24]; however, the main existence result is as follows.

Theorem 5.1 ([24]). *There is an economical SCCD$(v,3)$ for all $v\geq 3$.*

Proof. The proof is based on the following construction and several small examples. The authors show that if there is a SCCD$(v,3)$ with b blocks, then there is a SCCD$(v+4,3)$ with $2v+3+b$ blocks. Suppose the SCCD$(v,3)$ has blocks $B_1, B_2, \ldots, B_b$ and assume $B_b = \{1,2,3\}$. To obtain the new design with points $\{1,2,\ldots,v,a,b,c,d\}$, append the blocks

$$\{1,2,b\},\{2,b,d\},\{2,a,d\},\{2,a,c\},\{1,a,c\},\{1,c,d\},\{3,c,d\},$$
$$\{4,c,d\},\ldots,\{v,c,d\},\{v,b,c\},\{v,a,b\},\{v-1,a,b\},\ldots,\{3,a,b\}.$$

Combined with examples of economical SCCD$(v,3)$s for $v=3,4,5,6$, the construction allows for building economical SCCD$(v,3)$s for $v \geq 3$. □

The paper of Wallis, Yucas and Zhang also presents a variety of lower bounds on the number of blocks expected in SCCDs. Results are largely obtained by simple counting arguments or by presenting a reasonable SCCD with the given parameters.

In [22], van Rees gives a lower bound for the number of blocks in a SCCD $(3k-2,k)$.

Theorem 5.2 ([22]). *For $k \geq 4$, the minimum number of blocks in a SCCD $(3k-2,k)$ is $4k-1$.*

This result is proved by obtaining a contradiction to the assumption that a SCCD$(3k-2,k)$ with $4k-2$ blocks exists. We include the proof to illustrate the general methods associated with proving existence of SCCDs. Notice that proofs of existence for SCCDs are very different from those for configuration ordering (and we will see that they are also different from other minimal change results, including Gray codes and Ucycles). Other ordering existence results tend to be proved constructively, while existence of single-change designs is often proved by appealing to required properties.

Proof (Proof of Theorem 5.2). Recall that, for $i \geq 2$, a point x is said to be introduced in B_i if $x \in B_i \setminus B_{i-1}$. All points of B_1 are said to be introduced in B_1. Let T_i denote the set of points introduced i times and let $t_i = |T_i|$.

Assume that a SCCD$(3k-2,k)$ with $4k-2$ blocks exists. Such a SCCD is tight so there are no pairs covered more than once. If a point is introduced i times then it is in at least $i(k-1)$ pairs covered by the design. It is in an additional pair for every time it remains in a block. Since the number of pairs covered by the SCCD is $\binom{3k-2}{2}$, the tightness implies that $\binom{3k-2}{2} \geq i(k-1)$ and thus $t_i = 0$ for $i \geq 4$. Further, every point in T_3 must be dropped immediately from the block where it was introduced. If $t_3 > 1$, this makes it impossible for any pair of points in T_3 to be covered. Additionally, in order to avoid repeated pairs, the introductions of points from T_3 must be separated by at least $k-1$ blocks. That is, the indices of these blocks must differ by at least k. This implies that the largest block index possible for the first introduction of any $z \in T_3$ is $2k-2$.

What can be said about T_i for $i < 3$? Since there are more than $v-1$ blocks in the design, we know that $T_0 = \emptyset$. Any point $x \in T_1$ is introduced once and then must remain for exactly $2k-1$ blocks before being dropped which implies that the first block can contain at most one point from T_1. Finally, we consider points in T_2.

Suppose that a point from T_2 remains in r_1 blocks after its first introduction and in r_2 blocks after its second introduction before being dropped. Then $r_1 + r_2 = k+1$ and at least $k-1$ blocks must separate the first time the point is dropped from the second introduction.

Careful counting gives

$$b+k-1 = 5k-3 = t_1 + 2t_2 + 3t_3 \text{ and}$$
$$v = 3k-2 = t_1 + t_2 + t_3.$$

There are only two non-negative integer solutions to these equations: (1) $t_1 = k-1$, $t_2 = 2k-1, t_3 = 0$, and (2) $t_1 = k, t_2 = 2k-3, t_3 = 1$.

In both cases, each of the points from T_1 must appear only in a set of $2k-1$ consecutive blocks. If the first block contains a point of T_1 then the fact that all the pairs from T_1 must be covered means that the last introduction of a point from T_1 happens in B_{2k-1}, or earlier, and so the very last block that could contain any point from T_1 is B_{4k-3}. Thus, from the reversibility of SCCDs, we can assume that no point from T_1 is in the first block. Let us analyse each case.

Case 1: We can assume that B_1 is a subset of T_2. Since an element from T_2 cannot be in $k+1$ consecutive blocks, each of the points of B_1 must be dropped from the first k blocks and cannot be introduced again until B_{k+1} without creating a repeated pair. The lengths of runs of consecutive blocks (including B_1) that these appear in are $1,2,\ldots,k$. The number of consecutive blocks these appear in after their second introduction must be $k, k-1, \ldots, 2, 1$, respectively, and they have all appeared as pairs in the first block so none can appear together again. This means the number of blocks must be at least

$$k + (1+2+\cdots+k) = \frac{k^2+3k}{2},$$

which is greater than $4k-2$ when $k \geq 5$, a contradiction.

Now consider the case $k=4$. When $k=4$, $t_1 = 3$, $t_2 = 7$ and we assume an SCCD(10,4) with 14 blocks exists. As before, let $B_1 = \{1,2,3,4\}$ be a subset of T_2. Let $B_2 = \{1,2,3,5\}$. First, suppose $5 \in T_2$. Then B_2, B_3 and B_4 must drop $1,2$ and 3 in some order and B_5 must drop 5, otherwise one of these points would occur too often. After B_5, the points $1,2,3,4$ and 5 never occur together again, except for one block containing 4 and 5. So the number of blocks is at least $5+(4+3+2+1+1)-1 = 15$, a contradiction. Now suppose $5 \in T_1$. It must still be true that B_2, B_3 and B_4 drop $1,2$ and 3 in some order. Thus, B_5, B_6, B_7 and B_8 contain 5 but not $1,2$ or 3. No pairs of these points may appear together in the blocks following B_8; however, each must occur again in these last six blocks. But 4 cannot occur with $1,2$ or 3 after B_1, so 4 must occur in B_5, B_6, B_7 and B_8 also. But then B_8 has to drop two points—4 and 5—which contradicts the definition of single-change.

Table 5.1 A standardized tight SCCD$(12,4)$

1	1	1	1	1	1	1	1	1	7	7	7	7	8	8	8	8	8	8	5	5
2	2	2	2	8	9	9	11	11	11	11	11	11	11	11	11	12	12	12	12	12
3	3	3	7	7	7	10	10	10	10	10	10	10	10	10	9	9	9	9	9	6
4	5	6	6	6	6	6	6	12	12	3	4	5	5	2	2	2	3	4	4	4

Case 2: If the point $z \in T_3$ is not in B_1, then the same considerations as Case 1 imply that $T_1 \cap B_1 = \emptyset$ and $k = 4$. In the case of $k = 4$, the arguments from Case 1 leave only one possibility: that 5, which is introduced in B_2, is the single point in T_3. This point must be dropped immediately which will force one of $1, 2$ or 3 to remain in blocks until B_5, preventing it from being introduced a second time without covering a pair of points more than once. So we can assume that $z \in T_3 \cap B_1$. By reversal $z \in B_{4k-2}$ as well. Again we assume that $T_1 \cap B_1 = \emptyset$. We consider the pairs that z appears in. In B_1 it appears with $k-1$ points from T_2. In B_{4k-2} there are only $k-2$ points of T_2 remaining, so it must appear with all of these and precisely one point from T_1. Thus, we know that the reversed sequence must start with an element of T_1 in the first block. Counting the blocks (in the reversed sequence), we have

$$2k - 1 + 2 + (2 + 3 + \cdots + k - 1) = \frac{k^2 + 3k}{2} \leq 4k - 2,$$

which again implies that $k = 4$. If the point introduced in B_2 is from T_1, then it is impossible to place z two more times without introducing a repeated pair. Thus, the point introduced in B_2 is from T_2 and must be dropped from B_5. Further, this point must now be dropped immediately after its next appearance, which implies that its pairing with z must be obtained on B_5. B_5 can contain at most two members of T_1 and B_{14} must contain z and at least two members of T_1 (due to the properties of the reverse sequence). This forces both of these T_1 points to be dropped simultaneously in the reverse sequence, which contradicts the definition of single-change. □

In [17], Phillips and Preece enumerate and classify all standardized tight SCCD$(12,4)$s. This is the smallest possible v for tight SCCD$(v,4)$s and there are 2,554 of them. Table 5.1 presents a tight SCCD$(12,4)$ in the standardized form. Columns represent blocks. We note that in the literature, points that are unchanged from one block to the next are represented by a dot; however, in keeping with our representation of other orderings, we explicitly give all points of each block.

Phillips and Wallis define the notion of a **persistent pair** which is a pair of points that remain in blocks for a long run (defined in this paper to be $v - k$ in length) [18]. These are similar in concept to long bit runs in binary Gray codes [12]. Unfortunately, their existence makes forming a Ucycle (or sequential covering design) from a SCCD impossible. Phillips and Wallis use persistent pairs to computationally construct a tight SCCD$(12,4)$, tight SCCD$(13,4)$, tight SCCD$(15,4)$, and tight SCCD$(18,4)$. Additionally, they show that no triple can persist for more than $v - k - 2$ blocks in a tight SCCD(v,k) [18].

Table 5.2 A standardized tight SCCD$(12,4)$ with an outer expansion set of locations: $\{0,3,13,18\}$

1	1	1	1	1	1	1	1	1	4	4	4	3	6	2	2	2	2	3	3	4
2	2	2	7	7	9	10	10	12	12	12	12	12	12	12	12	12	7	7	6	6
3	3	6	6	8	8	8	11	11	11	11	11	11	11	11	11	10	10	10	10	10
4	5	5	5	5	5	5	5	5	5	7	8	8	8	8	9	9	9	9	9	9

In [20], Preece et al. define a **row regular** single-change design to be one where positions in the blocks are labelled and the number of changes are balanced across positions. That is, each row contains the same number of introduced points. This is similar to the balance required for balanced binary Gray codes (see Sect. 2.3.1.2). Preece et al. show there exists a row regular tight SCCD$(v,2)$ whenever $v \equiv 2,3 \pmod 4$. They show there cannot exist a row regular tight SCCD$(7,3)$, but that there do exist row regular tight SCCD$(10,3)$ and tight SCCD$(19,3)$. Furthermore, using a technique very similar to Theorem 5.1, they give a recursive construction which produces a row regular tight SCCD$(v+12,3)$ from a row regular tight SCCD$(v,3)$. This establishes that there are row regular tight SCCD$(v,3)s$ whenever $v \equiv 7,10 \pmod{12}$, $v \neq 7$ [20].

Preece et al. define some other internal properties which yield more general recursive constructions. If a tight SCCD(v,k) has a set of $v/(k-1)$ indices (otherwise known as locations) referring to blocks in the sequence, excluding the first and last, whose unchanged subsets (from the perspective of the preceding block) partition V, the set is called an **inner expansion set of locations**. If a tight SCCD(v,k) has a set of locations referring to blocks in the sequence, including at least one of the first and last, whose unchanged subsets (taking any $k-1$ subset of the first block) partition V, the set is called an **outer expansion set of locations**. Table 5.2 gives an example tight SCCD$(12,4)$ with an outer expansion set of locations. The set of unchanged subsets at positions $0,3,13$ and 18 is $\{\{2,3,4\},\{1,6,5\},\{12,11,8\},\{7,10,9\}\}$. Note further the long run of the pair $\{1,5\}$ from blocks one through eight, and the long run of the pair $\{11,12\}$ from blocks eight through fifteen.

Theorem 5.3 ([20]).

1. *If there exists a tight SCCD$(n(k-1),k)$ with either an inner or outer expansion set of locations, then there exists a tight SCCD$(n(k-1)$ $+1,k)$.*
2. *If there exists a standardized tight SCCD$(n(k-1),k)$ that has an outer expansion set of locations and there exists a standardized tight SCCD(v,k), then there exists a standardized tight SCCD$(v+(n-1)(k-1),k)$.*

Preece et al. are only able to construct the ingredients for these constructions for $k=4$, but using these, they prove that there exists a tight SCCD$(v,4)$ for $v \equiv 3,4 \pmod 9$ and $v \equiv 3 \pmod{12}$. Finally, they call for the search for a tight SCCD$(20,5)$ [20].

In [16], Phillips describes an algorithm which he uses to find tight SCCD$(20,5)$s. This backtrack search algorithm produced 35 such designs. Again, this is the smallest possible v for which a tight SCCD$(v,5)$ can exist. Further work on tight SCCDs (and other forms of single-change covering design) was undertaken by McSorley. In [15], McSorley presents a wide range of results, including existence of tight single change circular covering designs (SCCCDs).

Theorem 5.4 ([15]).

1. *A tight SCCCD$(v,2)$ exists for all $v \geq 3$.*
2. *A tight SCCCD$(v,3)$ exists for all $v \equiv 0,1 \pmod 4$, $v \geq 4$.*
3. *An economical SCCCD$(v,3)$ exists for all $v \equiv 2,3 \pmod 4$, $v \geq 6$.*

The SCCCDs in the theorem above are constructed by starting with a small design of the appropriate type, then adding a specified set of new blocks to obtain larger designs.

We give an example of the construction methods of McSorley. Arrange the elements of $[v-1]$ in order on a circle. McSorley defines the block A_i as follows: for $1 \leq i \leq v-1$, let A_i be the block containing i and the $k-1$ consecutive elements taken clockwise from the circle. Let $B_i = A_i \cup \{v\}$ and $\mathscr{B} = \{B_1, \ldots, B_{v-1}\}$.

Theorem 5.5 ([15]). *For $k \geq 3$ and $k+1 \leq v \leq 2k-2$, the blocks $\mathscr{B} = \{B_1, \ldots, B_{v-1}\}$, where $B_i = A_i \cup \{v\}$, form an economical SCCCD(v,k).*

Proof. We need only prove that $\mathscr{B}$ is a SCCCD(v,k) since it has the right number of blocks to be economical. The blocks of $\mathscr{B}$ clearly exhibit the required single-change property, that is, consecutive blocks differ by the removal of one element and the addition of one element. We now check that every pair is contained in at least one block of $\mathscr{B}$. Let $B_i = \{i, i+1, \ldots, i+k-2\} \cup \{v\}$, where addition is modulo $v-1$. Element $i+k-2$ is introduced in B_i, and thus, i is introduced in B_{i-k+2}. For $1 \leq i \leq v-1$, pair $\{i,v\}$ is covered by B_{i-k+2}, which accounts for all pairs that contain v. Thinking of B_i as a base block, we see that it contains all the differences $\{1,2,\ldots,k-2\}$ which is all differences in $\mathbb{Z}_{v-1}$ when $v \leq 2k-2$. Thus, every pair of points from $\mathbb{Z}_{v-1}$ will be contained in some translate of B_i.

Finally, we must ensure that no pair is in every block. Suppose $\{i,j\}$ is in every block. Then, without loss of generality, $1 \leq i \leq v-1$. But i is introduced in B_{i-k+2} and so cannot be in the previous block, a contradiction. □

For particular values of v, Theorem 5.5 yields tight designs.

Theorem 5.6 ([15]).

1. *For $k \geq 3$ and $v = 2k-2$, the blocks $\mathscr{B} = \{B_1, \ldots, B_{2k-3}\}$, where $B_i = A_i \cup \{2k-2\}$, form a tight SCCCD$(2k-2,k)$ having $2k-3$ blocks.*
2. *For $k \geq 2$ and $v = 2k-1$, the blocks $\mathscr{B} = \{B_1, \ldots, B_{2k-1}\}$, where $B_i = A_i \cup \{i+k-1\}$, form a tight SCCCD$(2k-1,k)$ having $2k-1$ blocks.*

Using methods similar to the proof of Theorem 5.5, McSorley constructs economical SCCCD(v,k)s for $v = 2k$.

Theorem 5.7 ([15]). *For $k \geq 2$ and $v = 2k$, there exists an economical SCCCD $(2k,k)$ having $2k+2$ blocks.*

Table 5.3 The number of blocks in economical SCND(v,k)s (smallest designs known to exist)

v \ k	4	5	6	7	8	9
5	4* (5)					
6	7 (7)	6* (7)				
7	10 (10)	9* (10)	9 (10)			
8	13* (13)	13 (13)	12* (13)	12 (13)		
9	17* (17)	17 (17)	16* (16)	16 (16)	15* (16)	
10	22 (22)	21* (21)	21 (21)	20* (20)	20 (20)	19* (21)

Table 5.4 A SCND(9,4) with 17 blocks

1	1	1	1	1	1	1	9	9	9	9	9	9	9	9	8	8
2	2	2	2	2	2	3	3	1	4	4	4	5	5	5	5	5
3	3	3	3	4	8	8	8	8	8	5	6	6	2	2	2	2
4	5	6	7	7	7	7	7	7	7	7	7	7	7	6	6	9

These designs are tight only for $k = 2$.

McSorley is able to construct a tight SCCCD$(9,4)$ and a tight SCCCD$(10,4)$. He also shows that if a tight SCCCD$(3k-3,k)$ or a tight SCCCD$(3k-2,k)$ exists, then $k = 2$ or 4. Finally, he explores the appropriate notions of isomorphism for these single-change designs and develops a notion of a canonical form [15].

In [8], the existence of single-change neighbour designs (SCNDs) is investigated. Constable et al. begin by investigating small cases. Table 5.3 lists the number of blocks in small economical SCNDs. The number of blocks in the smallest designs known to exist is given in brackets, and an asterisk indicates the bound is also tight.

Table 5.4 gives an example of a SCND. Columns represent blocks. Recall from the definition of SCNDs that cycles cover consecutive pairs, for example, the first block (cycle) in Table 5.4 covers pairs $\{1,2\},\{2,3\},\{3,4\},\{4,1\}$.

The main result of [8] is the determination of existence of SCND(v,k)s for $v = 4$.

Theorem 5.8 ([8]). *There exists an economical SCN$(v,4)$ for each $v \geq 6$.*

This result is proved using the following lemma and the known small economical SCNDs (constructed by computer).

Lemma 5.2 ([8]). *An economical SCND$(v,4)$ can be extended to an economical SCND$(v+8q,4)$, for all $q > 0$.*

An open question posed by Constable et al. is, does there exist a tight economical SCND$(k+1,k)$? None have been constructed and it has been shown than none exist for $k < 10$.

For sequential covering designs (SDs)—the Ucycle variant of single-change designs—Wallis has developed a number of constructions. Recall that the length of the sequence is denoted by t and is included in the design notation as SD(v,k,t). The smallest t for which a SD(v,k,t) exists is denoted by $g(v,k)$. Wallis first established the existence of shortest possible sequences for some small boundary cases.

Table 5.5 Sequential covering designs for some specific parameters

SD(8,3,18):	512634567853714827
SD(9,3,21):	316427531283495687912
SD(10,3,30):	164275312834956879120365407890
SD(10,4,21):	671238456789063194520
SD(12,5,24):	8912340567890ab7312a564b
SD(14,6,30):	4590a12345b67890abcd81236c745d

Lemma 5.3 ([25]).

1. *When $k \le v < 2k$, a SD$(v,k,2v-k)$ exists.*
2. *When v is odd, a SD$(v,2,1+v(v-1)/2)$ exists.*
3. *When v is even, a SD$(v,2,v^2/2)$ exists.*

He then examined some recursive constructions.

Theorem 5.9 ([25]). *Suppose there exists a SD(v,k,t), then there exists a SD$(v+1,k,t_1)$ and a SD$(v+2,k,t_2)$ with*

$$t_1 = t+v+\lceil v/(2k-2)\rceil - k+1, \text{ and}$$

$$t_2 = t+v+\lceil (v+1)/(k-1)\rceil - k+3.$$

When these are combined and considered with the bound from Theorem 3.1 (see page 69) we get $g(2k,k) = 4k$, $g(2k+1,k) = 4k+2$ and $4k+4 \le g(2k+2,k) \le 5k+1$. Wallis was also able to construct some other sequences for specific values; these are given in Table 5.5. Note that they are not necessarily optimal.

5.2 Gray Codes and Universal Cycles for TTSs

Twofold triple systems are a natural family with which to begin our investigation of Gray codes and Ucycles because both the index and the strength of the design are one less than the block size. Furthermore, the small size of the blocks makes it easy to look directly at the points in each block of the design. A minimal change ordering for a TTS is a listing of the blocks of the design such that consecutive blocks differ in one entry. Such a minimal change ordering can be realized in two ways. As a Gray code, this is directly analogous to minimal change combinatorial Gray codes for k-subsets of $[n]$ and single-change covering designs. As a rank three Ucycle, this is similar in appearance to a Ucycle for k-subsets of $[n]$; however, the construction methods are very different.

As noted in Chap. 2, a TTS(v) exists if and only if $v \equiv 0,1 \pmod 3$. Lemma 3.5 (page 73) implies that if there exists a rank three Ucycle for a TTS(v), then three must divide r, where $r = v-1$ in this case. This implies that v must be equivalent to one modulo three; therefore, $v \equiv 1,4,7,10 \pmod{12}$ are the only values of v for which a TTS(v) may admit a Ucycle of rank three. For all $v \equiv 0,1 \pmod 3$, a 2-intersecting Gray code for a TTS(v) may exist. In this section, we prove that for $v \equiv 1,3,4,7 \pmod{12}$, with sporadic exceptions, there exists a TTS(v) that admits a 2-intersecting cyclic Gray code for its blocks. When $v \equiv 1,4,7 \pmod{12}$, this ordering can be written as a Ucycle of rank three. Determining the existence of rank three Ucycles for TTS(v)s with $v \equiv 10 \pmod{12}$, and the existence of 2-intersecting cyclic Gray codes for TTS(v)s with $v \equiv 0,6,9 \pmod{12}$, remains an open problem.

The existence of a 2-intersecting cyclic Gray code for the blocks of a TTS is equivalent to the existence of a Hamilton cycle in the 2 block-intersection graph of the design. It is also equivalent to saying that the blocks admit an A_3-ordering (see Fig. 1.1 on page 4). In [6], M. Colbourn and Johnstone construct a TTS having a 2 block-intersection graph that does not admit a Hamilton cycle. In proving this result, M. Colbourn and Johnstone first establish the following lemma.

Lemma 5.4 ([6]). *Given a TTS(v), each component of the corresponding 2 block-intersection graph is 3-connected.*

Proof. Suppose that one of the connected components has a cutset of size less than three. We break the analysis into two cases based on cutset size.

Case 1: Suppose the cutset consists of a single vertex $A = \{a,b,c\}$. In a slight abuse of terminology, we say this block (really this vertex) is adjacent to three distinct blocks: $\{a,b,x\}$, $\{a,c,y\}$ and $\{b,c,z\}$. Since every pair of points appears precisely twice in the design, the blocks $\{a,b,x\}$, $\{a,b,c\}$ and $\{a,c,y\}$ lie on a cycle consisting of vertices representing blocks that contain point a. That is, the subgraph induced by the blocks containing a is a union of cycles. Similarly, $\{a,b,x\}$, $\{a,b,c\}$ and $\{b,c,z\}$ lie on a cycle in which every vertex corresponds to a block containing b, and the same is true for the blocks containing c. Therefore, there are three cycles through vertex A which are edge disjoint except for the three edges incident to A. Thus, the neighbours of A (and all other vertices of the graph) are connected by paths that avoid A, so removing A cannot disconnect this component, a contradiction.

Case 2: Now suppose that the cutset has cardinality two. If the two vertices of the cutset are adjacent, say $A = \{a,b,c\}$ and $Z = \{b,c,z\}$, then the cycles from the previous argument show that the neighbours of A are connected to each other and to the neighbours of Z by paths which avoid both A and Z. The same statement holds for the neighbours of Z; therefore, the two vertices of the cutset must not be adjacent.

Let A and B be vertices of a cutset that are non-adjacent. Let the neighbours of A be X, Y and Z. From Case 1 we know there are edge disjoint paths connecting each pair of vertices in $\{X,Y,Z\}$. Since B can appear on at most one of these paths, the removal of B cannot disconnect any of the neighbours of A. The same argument holds for vertex B. Thus, there can be no cutset of size two. □

The main result of [6] is the determination of a TTS which has no minimal change presentation, that is, has no Hamilton cycle in the 2 block-intersection graph of the design. An obvious condition for existence of a Hamilton cycle in a graph is that the graph must be connected; therefore, M. Colbourn and Johnstone deal only

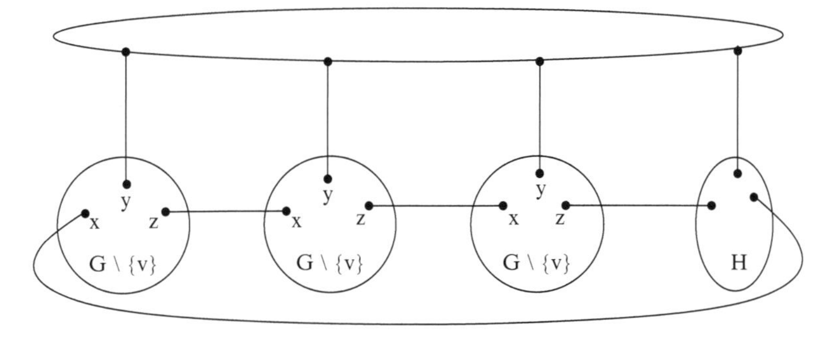

Fig. 5.1 The 2 block-intersection graph of a TTS having no Hamilton cycle

with TTSs that are simple. By this, they mean that the TTS has no repeated blocks *and* the TTS has no non-trivial sub-design. Using an approach due to McCuaig, M. Colbourn and Johnstone present a construction for an infinite family of graphs which can be viewed as 2 block-intersection graphs of TTSs. Let $G = (V,E)$ be a cubic, 3-connected graph that does not contain a Hamilton cycle. A good example of such a graph is the Petersen graph. Let $v \in V$ and let the three neighbours of v be $x,y,z \in V$. Create three copies of $G \setminus \{v\}$ and connect them as shown in Fig. 5.1, then add any graph H such that the new graph created can be labelled as the 2 block-intersection graph of some TTS. This new graph does not contain a Hamilton cycle. The authors claim that such an H exists; however, they give only one example, choosing the Petersen graph for G and choosing H such that the resulting graph can be labelled as the 2 block-intersection graph of a TTS(19). The authors do not prove or discuss whether it is possible to find other graphs H which complete the graph in Fig. 5.1 (with G the Petersen graph) to the 2 block-intersection graph of a TTS, nor do they indicate whether, given an arbitrary G with the required properties, it is always possible to find an H which completes the copies of G to the 2 block-intersection graph of some TTS. The existence of H is worth investigating as M. Colbourn and Johnstone note that there exist many cubic 3-connected graphs which have no corresponding TTS or partial TTS labelling.

While M. Colbourn and Johnstone prove the existence of a TTS having no minimal change presentation, they predict that for all admissible v there *do* exist minimal change TTS(v)s. Dewar proved that there exists a TTS(v), for each $v \equiv 1,3,4,7 \pmod{12}$, with sporadic exceptions, that admits a minimal change presentation. Further, she proved that for $v \equiv 1,4,7 \pmod{12}$, the stronger Ucycle representation exists [9]. The existence of a minimal change ordering for at least one TTS of each admissible order is perhaps the best general result we can hope for, as M. Colbourn and Johnstone's work implies that we will not be able to prove that all TTS(v)s admit minimal change orderings.

The existence of cyclic Gray codes and Ucycles for the blocks of twofold triple systems is proved in very similar ways. In fact, the way to think of cyclic Gray codes is as Ucycles that cannot be completed unless the change rule is relaxed in some places. Define a **partial Ustring** to be a sequence having Ucycle properties (although possibly not cyclic) that represents some, but not all, blocks of a design. To create a Ucycle for a design, we join partial Ustrings, which together represent all blocks of the design, so that the Ucycle property holds across joins. To create a cyclic Gray code for a design, we take collections of partial Ustrings, which together contain all blocks of the design, and join them so that the minimal change property holds across joins.

Suppose U is a partial Ustring containing some blocks of a TTS(v), $S=(V,\mathscr{B})$. What properties must U have? First, the list of blocks induced by U will have the property that the intersection of four consecutive blocks is empty and the intersection of three consecutive blocks is exactly one. This is due to the fact that consecutive blocks are "picked off" U by shifting right by one entry. For example, suppose v,w,x,y,z is a partial Ustring. The blocks represented by this partial Ustring are $\{v,w,x\}$, $\{w,x,y\}$ and $\{x,y,z\}$. We cannot have three consecutive blocks intersecting in more than one point because every pair of points appears in exactly two blocks of a TTS. On the other hand, every point in U appears in exactly three blocks unless U is not cyclic (in which case the endpoints of U appear in fewer blocks). Second, a complete Ucycle must have the property that all pairs $\{x,y\}\in V\times V$ either appear adjacent in the Ucycle exactly once (either as x,y or y,x) or appear exactly twice at distance two (either as $x,\_,y$ or $y,\_,x$). Consider the partial Ustring in the example above. In addition to the block $\{w,x,y\}$ there is exactly one other block in $\mathscr{B}$ containing the pair $\{w,y\}$. The other block must appear in the Ucycle as $w,\_,y$ or $y,\_,w$, because a consecutive occurrence of w and y would result in two additional blocks containing this pair. These two properties govern the creation of partial Ustrings for TTSs.

For each admissible order v, there exists a TTS(v) that is either cyclic or 1-rotational. Recall that a cyclic design of order v has an automorphism $\pi\colon i\longmapsto i+1 \pmod{v}$. A 1-rotational design of order v has an automorphism with one fixed point and a cycle of length $v-1$, where the fixed point is denoted ∞ and the automorphism is $\pi\colon i\longmapsto i+1 \pmod{v-1}$. Cyclic TTS$(v)$s exist for all $v\equiv 0,1,3,4,7,9 \pmod{12}$, except $v=9$, while 1-rotational TTS(v)s exist for all $v\equiv 0,1 \pmod{3}$. In [9], Dewar uses the structural properties of such designs to construct partial Ustrings.

Recall that a supplementary difference set, $\mathscr{S}$, for a cyclic BIBD(v,k,λ) must have the following properties: (1) $|S|=k$, for all $S\in\mathscr{S}$, and (2) the collection $\{x-y: x,y\in S, S\in\mathscr{S}, x\neq y\}$ must contain every element of $\{1,2,\ldots,v\}$ exactly λ times. Each set $S\in\mathscr{S}$ yields a base block, and when all base blocks are developed through $\mathbb{Z}_v$ they generate the design. For a 1-rotational BIBD(v,k,λ), the collection of base blocks must each be of size k and together must contain the differences $\{1,2,\ldots,v-2\}$ exactly λ times. Note that while the point ∞ will be in some base blocks, the differences involving this fixed point are ignored. Given a collection of blocks, these two difference properties can be used to determine if the blocks form a set of base blocks for a cyclic or 1-rotational design.

The differences associated to the base blocks of a cyclic BIBD(v,k,λ) reside in $\mathbb{Z}_v \setminus \{0\}$. When determining the appearance of each of these differences in the base blocks of a design, we will work with the set

$$\{-(\lceil v/2 \rceil - 1), -(\lceil v/2 \rceil - 2), \ldots, \lfloor v/2 \rfloor\}$$

which is equivalent to $\mathbb{Z}_v$. Notice that if the difference $x-y$ appears in a base block then so does the difference $v-(x-y) \equiv y-x \pmod{v}$, therefore, we need only ensure that each difference $d \in \{1,2,\ldots,\lfloor v/2 \rfloor\}$ appears λ times in the base blocks of a cyclic BIBD(v,k,λ). In fact, the appearance of d or $-d$, for $d \in \{1,2,\ldots,\lfloor v/2 \rfloor\}$, in a base block contributes to the count of the total number of times the difference d appears in the base blocks in question. Similarly, the differences associated to the base blocks of a 1-rotational BIBD(v,k,λ) reside in $\mathbb{Z}_{v-1} \setminus \{0\}$. In this case, we need only ensure that each difference $d \in \{1,2,\ldots,\lfloor (v-1)/2 \rfloor\}$ appears λ times in the base blocks of a 1-rotational BIBD(v,k,λ).

To prove the existence of rank three Ucycles and 2-intersecting cyclic Gray codes for certain TTS(v)s, we will first construct partial Ustrings that contain all base blocks of the TTS in question. To construct partial Ustrings we must define a new structure and introduce some new notation. Let $D_v = d_0, d_1, \ldots, d_{\ell-1}$ be a sequence of elements from $\mathbb{Z}_v \setminus \{0\}$ such that $d_i \neq d_j \pmod{v}$ when $i \neq j$. Let σ_{D_v} be the sequence of sums of pairs of consecutive elements in D_v, that is, $\sigma_{D_v} = d_0 + d_1 \pmod{v}, d_1 + d_2 \pmod{v}, \ldots, d_{\ell-2} + d_{\ell-1} \pmod{v}$. When this sequence of sums includes the sum of the last and first elements in D_v denote this sequence $\sigma'_{D_v} = \sigma_{D_v}, d_{\ell-1} + d_0 \pmod{v}$. For the remainder of this section, all operations involving elements of D_v will be assumed to be done modulo v. Given a sequence S, let $\{S\}$ denote the collection of elements appearing in S (including repeated elements) and let $\pm\{S\}$ denote the collection of elements in S and the negative of these elements. Let $x\{S\}$, for $x \in \mathbb{Z}^+$, denote the collection of elements appearing in S, with each element repeated x times.

Definition 5.1 (cyclic difference sequence of order v). Let $D_v = d_0, d_1, \ldots, d_{\ell-1}$ be a sequence of elements in $\mathbb{Z}_v \setminus \{0\}$ such that $d_i \neq d_j \pmod{v}$ when $i \neq j$:

- If $v \equiv 0 \pmod{2}$, then D_v is a cyclic difference sequence of order v if and only if $d_i \neq v/2$, for all $i \in \{0,1,\ldots,\ell-1\}$, and $2\{(\mathbb{Z}_v \setminus \{0\}) \setminus \pm\{D_v\}\} = \pm\{\sigma'_{D_v}\}$.
- If $v \equiv 1 \pmod{2}$, then D_v is a cyclic difference sequence of order v if and only if $2\{(\mathbb{Z}_v \setminus \{0\}) \setminus \pm\{D_v\}\} = \pm\{\sigma'_{D_v}\}$.

This definition is equivalent to saying that every non-zero element or its negative (but not both) in $\{1,2,\ldots,\lfloor v/2 \rfloor\}$ must appear once in D_v or must appear twice in σ'_{D_v}. Note that when $v \equiv 0 \pmod{2}$, the difference $v/2 \equiv -v/2 \pmod{v}$; thus, the difference $v/2$ must appear in σ'_{D_v} and must appear exactly once. Given its properties, a cyclic difference sequence must be of length $(v-1)/3$. To illustrate the concept, we give examples of each type of cyclic difference sequence. The sequence $D_{16} = 1,4,2,3,7$ is a cyclic difference sequence of order 16 with $\sigma'_{D_{16}} = 5,6,5,-6,8$. The sequence $D_{25} = 5,-6,7,-10,12,-8,11,-9$ is a cyclic difference sequence of order 25 with $\sigma'_{D_{25}} = -1,1,-3,2,4,3,2,-4$.

Definition 5.2 (difference sequence of order v). Let $D_v = d_0, d_1, \ldots, d_{\ell-1}$ be a sequence of elements in $\mathbb{Z}_v \setminus \{0\}$ such that $d_i \neq d_j \pmod{v}$ when $i \neq j$:

- If $v \equiv 0 \pmod{2}$, then D_v is a difference sequence of order v if and only if $d_i \neq v/2$, for all $i \in \{0, 1, \ldots, \ell-1\}$, and $2\{(\mathbb{Z}_v \setminus \{0\}) \setminus \pm\{D_v\}\} \cup \pm\{d_{\ell-1}\} = \pm\{\sigma_{D_v}\}$.
- If $v \equiv 1 \pmod{2}$, then D_v is a difference sequence of order v if and only if $2\{(\mathbb{Z}_v \setminus \{0\}) \setminus \pm\{D_v\}\} \cup \pm\{d_{\ell-1}\} = \pm\{\sigma_{D_v}\}$.

This definition is equivalent to saying that every non-zero element or its negative (but not both) in $\{1, 2, \ldots, \lfloor v/2 \rfloor\}$ must appear once in D_v or must appear twice in σ_{D_v}, except for the last element in D_v which must appear exactly once in D_v and exactly once in σ_{D_v}. In order to have these properties, a difference sequence must be of length $(v+1)/3$. To illustrate the concept, we give examples of each type of difference sequence. The sequence $D_{14} = 3, -1, -6, 4, 5$ is a difference sequence of order 14 with $\sigma_{D_{14}} = 2, 7, -2, -5$. Notice that 5 appears both in D_{14} and in $\sigma_{D_{14}}$ (as -5) and that 7 appears exactly once in $\sigma_{D_{14}}$. The sequence $D_{11} = 1, 3, 2, 4$ is a difference sequence of order 11 with $\sigma_{D_{11}} = 4, 5, -5$.

Difference sequences and cyclic difference sequences may seem like artificial constructs; however, their properties reflect their intended use: cyclic and non-cyclic difference sequences can be used to construct the base blocks of twofold triple systems. We will describe the construction in a following paragraph; however, to motivate the definitions of difference sequences, we note that the construction method will imply that if D_v is cyclic, each entry of D_v will appear as a difference in two base blocks, while each entry of σ'_{D_v} will appear once in a base block. If D_v is not cyclic, each entry of D_v will appear as a difference in two base blocks, except the last which will appear in one base block, and each entry of σ_{D_v} will appear once in a base block.

Independently, Aldred, Bailey, McKay and Wanless [1] have defined a concept similar to a difference sequence. Given a positive integer v, they look for sequences of length $v-1$ containing elements of $\mathbb{Z}_v \setminus \{0\}$ such that the following properties hold: (1) the sum of each pair of consecutive elements in the sequence is distinct and non-zero, (2) the non-zero element missing from these sums is 1 if v is odd and $v/2+1$ if v is even, and (3) the sum of the last and the first elements in the sequence is one. Aldred, Bailey, McKay and Wanless have also defined several variants of this sequence—the interested reader is directed to [1] for further details.

The existence of difference sequences is an interesting question in and of itself; however, the concept of a difference sequence was developed because of its implication for designs. Given a cyclic difference sequence of order v, $D_v = d_0, d_1, \ldots, d_{\ell-1}$, define the non-cyclic sequence

$$U'_{D_v} = 0, d_0, d_0 + d_1, d_0 + d_1 + d_2, \ldots, \sum_{i=0}^{\ell-1} d_i, \sum_{i=0}^{\ell-1} d_i + d_0,$$

where each sum is taken modulo v. Similarly, given a difference sequence of order v, $D_v = d_0, d_1, \ldots, d_{\ell-1}$, define the non-cyclic sequence

$$U_{D_v} = \infty, 0, d_0, d_0 + d_1, d_0 + d_1 + d_2, \ldots, \sum_{i=0}^{\ell-1} d_i,$$

where each sum is taken modulo v.

Lemma 5.5 ([9]).

1. *The existence of a cyclic difference sequence of order v implies the existence of a cyclic TTS(v) and the existence of a partial Ustring representing its base blocks.*
2. *The existence of a difference sequence of order $v-1$ implies the existence of a 1-rotational TTS(v) and the existence of a partial Ustring representing its base blocks.*

Proof. (1) Let $D_v = d_0, d_1, \ldots, d_{\ell-1}$, $\ell = (v-1)/3$, be a cyclic difference sequence of order v. Construct the sequence

$$U'_{D_v} = 0, d_0, d_0 + d_1, d_0 + d_1 + d_2, \ldots, \sum_{i=0}^{\ell-1} d_i, \sum_{i=0}^{\ell-1} d_i + d_0,$$

where each sum is taken modulo v. We prove this sequence is a non-cyclic partial Ustring representing the base blocks of a cyclic TTS(v). The elements of U'_{D_v} are points in $\mathbb{Z}_v$ and every set of three consecutive elements represents a block; therefore, U'_{D_v} represents an ordering of ℓ blocks. Each of the differences in D_v appears in two of the blocks represented by U'_{D_v} and each of the differences in σ'_{D_v} appears in one of the blocks represented by U'_{D_v}. To see this, consider a portion of U'_{D_v}, say

$$\sum_{i=0}^{j-1} d_i, \sum_{i=0}^{j} d_i, \sum_{i=0}^{j+1} d_i, \sum_{i=0}^{j+2} d_i.$$

This sequence represents the blocks

$$\left\{\sum_{i=0}^{j-1} d_i, \sum_{i=0}^{j} d_i, \sum_{i=0}^{j+1} d_i\right\} \text{ and } \left\{\sum_{i=0}^{j} d_i, \sum_{i=0}^{j+1} d_i, \sum_{i=0}^{j+2} d_i\right\}.$$

The differences d_j, d_{j+1} and $d_j + d_{j+1}$ appear in the first block and the differences d_{j+1}, d_{j+2} and $d_{j+1} + d_{j+2}$ appear in the second block. In general, each element of D_v is a difference in two consecutive blocks, whereas the sum of two consecutive elements in D_v is a difference in one block. The structure of D_v and σ'_{D_v} implies that the set of blocks represented by U'_{D_v} has each difference in $\mathbb{Z}_v \setminus \{0\}$ appearing exactly twice in a block. Therefore, the partial Ustring U'_{D_v} represents the base blocks of a cyclic TTS(v).

(2) Let $D_{v-1} = d_0, d_1, \ldots, d_{\ell-1}$, $\ell = v/3$, be a difference sequence of order $v-1$. Construct the sequence

$$U_{D_{v-1}} = \infty, 0, d_0, d_0 + d_1, d_0 + d_1 + d_2, \ldots, \sum_{i=0}^{\ell-1} d_i,$$

where each sum is taken modulo $v-1$. We prove the sequence $U_{D_{v-1}}$ is a non-cyclic partial Ustring representing the base blocks of a 1-rotational TTS(v). Each of the differences in D_{v-1}, except $d_{\ell-1}$, appears in two of the blocks represented by $U_{D_{v-1}}$ and each of the differences in $\sigma_{D_{v-1}}$ appears in one of the blocks represented by $U_{D_{v-1}}$. By definition, $d_{\ell-1}$ appears in $\sigma_{D_{v-1}}$ exactly once and in D_{v-1} exactly once; therefore, each difference in $\mathbb{Z}_{v-1} \setminus \{0\}$ appears exactly twice in a block. The partial Ustring $U_{D_{v-1}}$ represents the base blocks of a 1-rotational TTS(v). □

The cyclic difference sequence of order 25, $D_{25} = 5, -6, 7, -10, 12, -8, 11, -9$, induces the non-cyclic partial Ustring $U'_{D_{25}} = 0, 5, -1, 6, -4, 8, 0, 11, 2, 7$. This partial Ustring represents the list of blocks $\{0,5,24\}$, $\{5,24,6\}$, $\{24,6,21\}$, $\{6,21,8\}$, $\{21,8,0\}$, $\{8,0,11\}$, $\{0,11,2\}$ and $\{11,2,7\}$. These are the base blocks of a cyclic TTS(25) as each difference in $\mathbb{Z}_{25} \setminus \{0\}$ occurs twice. The difference sequence $D_{14} = 3, -1, -6, 4, 5$ of order 14 induces a non-cyclic partial Ustring $U_{D_{14}} = \infty, 0, 3, 2, -4, 0, 5$. This partial Ustring represents the list of blocks $\{\infty, 0, 3\}$, $\{0,3,2\}$, $\{3,2,10\}$, $\{2,10,0\}$, and $\{10,0,5\}$. These are the base blocks of a 1-rotational TTS(15) as each difference in $\mathbb{Z}_{14} \setminus \{0\}$ occurs twice.

We refer again to the work of Aldred, Bailey, McKay and Wanless [1]. Their sequence (described just before Lemma 5.5) can be developed in a similar way to that described in the proof of Lemma 5.5 to obtain a circular sequence of length $v(v-1)$ on v elements in which every ordered pair of distinct elements occurs exactly once as neighbours and exactly once with a single item between them.

Lemma 5.6 ([9]). *Let D_v be a cyclic difference sequence of order v. If the sum of the elements in D_v is coprime to v, then there exists a Ucycle of rank three for the blocks of the cyclic TTS(v) induced by D_v.*

Proof. Given D_v—a cyclic difference sequence of order v—and U'_{D_v}—the non-cyclic partial Ustring induced by D_v—create $v-1$ other non-cyclic partial Ustrings by developing U'_{D_v} through $\mathbb{Z}_v \setminus \{0\}$. As U'_{D_v} represents the base blocks of a TTS(v), this collection of partial Ustrings represents all the blocks of the TTS(v). It remains to join these sequences together to form a Ucycle for the design. The first block represented by U'_{D_v} is $\{0, d_0, d_0 + d_1\}$ and the last block represented by U'_{D_v} is $\{\sum_{i=0}^{\ell-2} d_i, \sum_{i=0}^{\ell-1} d_i, \sum_{i=0}^{\ell-1} d_i + d_0\}$, with points appearing in this order in U'_{D_v}. In order to maintain the Ucycle representation across joins, we must join the end of U'_{D_v} to another partial Ustring which must begin with $\sum_{i=0}^{\ell-1} d_i, \sum_{i=0}^{\ell-1} d_i + d_0, x$, for some $x \in \mathbb{Z}_v$. Notice that the sequence $U'_{D_v} + \sum_{i=0}^{\ell-1} d_i$ has first block $\{\sum_{i=0}^{\ell-1} d_i, \sum_{i=0}^{\ell-1} d_i + d_0, \sum_{i=0}^{\ell-1} d_i + d_0 + d_1\}$. We can join the end of U'_{D_v} to the beginning of $U'_{D_v} + \sum_{i=0}^{\ell-1} d_i$, then we can join the end of $U'_{D_v} + \sum_{i=0}^{\ell-1} d_i$ to the beginning of $U'_{D_v} + (2 \cdot \sum_{i=0}^{\ell-1} d_i)$, and so on. Note that in joining partial Ustrings together we identify their two shared points. As long as $\sum_{i=0}^{\ell-1} d_i$ is coprime to v (the number of partial Ustrings in question) all partial Ustrings can be joined to form a single Ucycle for the design. □

Recall that the cyclic difference sequence $D_{25} = 5,-6,7,-10,\ 12,-8,11,-9$ induces the non-cyclic partial Ustring $U'_{D_{25}} = 0,5,-1,6,-4,8,0,11,2,7$. The sum of the entries in D_{25} is $2 \pmod{25}$. Since 2 and 25 are coprime, we can create a Ucycle for the TTS(25) having base blocks represented by $U'_{D_{25}}$. In the sequence below, the pairs of bold entries indicate the join of partial Ustrings (the bold points appear both at the end of one partial Ustring and at the beginning of another partial Ustring):

$$\mathbf{0},\mathbf{5},-1,6,-4,8,0,11,\mathbf{2},\mathbf{7},1,8,-2,10,2,-12,$$

$$\mathbf{4},\mathbf{9},3,10,0,12,4,-10,\mathbf{6},\mathbf{11},5,12,2,-11,6,-8,$$

$$\mathbf{8},\mathbf{-12},7,-11,4,-9,8,-6,\mathbf{10},\mathbf{-10},9,-9,6,-7,10,-4,$$

$$\mathbf{12},\mathbf{-8},11,-7,8,-5,12,-2,\mathbf{-11},\mathbf{-6},-12,-5,10,-3,-11,0,$$

$$\mathbf{-9},\mathbf{-4},-10,-3,12,-1,-9,2,\mathbf{-7},\mathbf{-2},-8,-1,-11,1,-7,4,$$

$$\mathbf{-5},\mathbf{0},-6,1,-9,3,-5,6,\mathbf{-3},\mathbf{2},-4,3,-7,5,-3,8,$$

$$\mathbf{-1},\mathbf{4},-2,5,-5,7,-1,10,\mathbf{1},\mathbf{6},0,7,-3,9,1,12,$$

$$\mathbf{3},\mathbf{8},2,9,-1,11,3,-11,\mathbf{5},\mathbf{10},4,11,1,-12,5,-9,$$

$$\mathbf{7},\mathbf{12},6,-12,3,-10,7,-7,\mathbf{9},\mathbf{-11},8,-10,5,-8,9,-5,$$

$$\mathbf{11},\mathbf{-9},10,-8,7,-6,11,-3,\mathbf{-12},\mathbf{-7},12,-6,9,-4,-12,-1,$$

$$\mathbf{-10},\mathbf{-5},-11,-4,11,-2,-10,1,\mathbf{-8},\mathbf{-3},-11,-2,-12,0,-8,3,$$

$$\mathbf{-6},\mathbf{-1},-7,0,-10,2,-6,5,\mathbf{-4},\mathbf{1},-7,2,-8,4,-4,7,$$

$$\mathbf{-2},\mathbf{3},-3,4,-6,6,-2,9,\mathbf{0},\mathbf{5},-3,6,-4,8,0,11,$$

$$\mathbf{2},\mathbf{7},1,7,-2,10,2,-11,\mathbf{4},\mathbf{9},1,10,0,12,4,-10,$$

$$\mathbf{6},\mathbf{11},5,11,2,-11,6,-7,\mathbf{8},\mathbf{-12},5,-11,4,-9,8,-6,$$

$$\mathbf{10},\mathbf{-10},9,-10,6,-7,10,-3,\mathbf{12},\mathbf{-8},9,-7,8,-5,12,-2,$$

$$\mathbf{-11},\mathbf{-6},-12,-6,10,-3,-11,1,\mathbf{-9},\mathbf{-4},-12,-3,12,-1,-9,2,$$

$$\mathbf{-7},\mathbf{-2},-8,-2,-11,1,-7,5,\mathbf{-5},\mathbf{0},-8,1,-9,-5,6,$$

$$\mathbf{-3},\mathbf{2},-4,2,-7,5,-3,9,\mathbf{-1},\mathbf{4},-4,5,-5,-1,10,$$

$$\mathbf{1},\mathbf{6},0,6,-3,9,1,-12,\mathbf{3},\mathbf{8},0,9,-1,3,-11,$$

$$\mathbf{5},\mathbf{10},4,10,1,-12,5,-8,\mathbf{7},\mathbf{12},4,-12,3,7,-7,$$
$$\mathbf{9},-\mathbf{11},8,-11,5,-8,9,-4,\mathbf{11},-\mathbf{9},8,-8,7,11,-3,$$
$$-\mathbf{12},-\mathbf{7},12,-7,9,-4,-12,0,-\mathbf{10},-\mathbf{5},12,-4,11,-10,1,$$
$$-\mathbf{8},-\mathbf{3},-9,-3,-12,0,-8,4,-\mathbf{6},-\mathbf{1},-9,0,-12,-6,5,$$
$$-\mathbf{4},\mathbf{1},-5,1,-8,4,-4,8,-\mathbf{2},\mathbf{3},-5,4,-8,-2,9.$$

Theorem 5.10 ([9]).

1. *For each* $v \equiv 1 \pmod{12}$, *with* $v \not\equiv 0 \pmod{5}$, *there exists a* TTS(v) *that admits a Ucycle of rank three.*
2. *For each* $v \equiv 4 \pmod{12}$, *with* $v \not\equiv 0 \pmod{5}$, *there exists a* TTS(v) *that admits a Ucycle of rank three.*
3. *For each* $v \equiv 7 \pmod{12}$, *there exists a* TTS(v) *that admits a Ucycle of rank three.*

Proof. To prove the existence of Ucycles for TTS(v)s with $v \equiv 1,4,7 \pmod{12}$, we present a cyclic difference sequence for each equivalence class. We show that Lemma 5.6 holds for case (1) and leave the work for the remaining cases as exercises:

(1) Suppose $v \equiv 1 \pmod{12}$ and let $v = 12m+1$. Define sequences

$$\alpha = -(3m+1), 3m, -(3m+2), 3m-1, \ldots, 2m+1, -(4m+1) \text{ and}$$
$$\beta = 4m+2, 6m, -(4m+3), 6m-1, -(4m+4), 6m-2, \ldots$$
$$\ldots, -(5m-1), 5m+3, -5m, 5m+2, 5m+1,$$

and set $D_v = \alpha, \beta$. Notice that in each sequence alternate elements form subsequences in which entries progress by plus one or minus one, except the first and last entries of β. Ignoring signs, the union of the elements in α and β is the set $\{2m+1, \ldots, 6m\} \subset \mathbb{Z}_v$. As $\sigma_\alpha = -1, -2, \ldots, -2m$ and $\sigma_\beta = 2m-1, 2m-3, 2m-4, \ldots, 3, 2, -(2m-2)$, it remains to determine if the differences 1 and $2m$ (or the negation of these values) appear once more as sums of two consecutive elements in D_v. At the join of the sequences α and β we have $\ldots, -4m-1, 4m+2, \ldots$, which yields the sum 1. Because D_v is a cyclic sequence we also consider the sum of the last entry in β and the first entry in α. This sum is $5m+1-3m-1 = 2m$.

It remains to determine when the sum of the entries in D_v is coprime to v. The sum of the entries in D_v can be obtained by taking the sum of the first and every alternate value in σ_{D_v}, plus the last element in D_v if it is not taken into account by these sums. From σ_α we have the sum of the odd values from -1 to $-(2m-1)$. The sum of differences across the join of α and β is one. Finally, taking the second element in σ_β and every alternate element, yields a sum of the positive odd values from 3 to $2m-3$ plus $10m+3$ (the sum of $5m+2$ and $5m+1$). Therefore, the sum of the entries in D_v is $8m+4$. Since $\gcd(8m+4,12m+1)=\gcd(10,4m-3)$, we conclude that $\gcd(8m+4,12m+1)=1$ for all $m \not\equiv 2 \pmod 5$. Thus, the sum of the elements in D_v is coprime to v for all $v \not\equiv 0 \pmod 5$, which implies that there exists a cyclic TTS(v) admitting a Ucycle of rank three for $v \equiv 1,13,37,49 \pmod{60}$.

(2) Suppose $v \equiv 4 \pmod{12}$ and let $v=12m+4$. Define sequences

$$\alpha = 4m+1,-(6m+1),4m+2,-6m,\ldots,-(5m+2),5m+1, \text{ and}$$
$$\beta = -(3m+1),3m,-(3m+2),3m-1,\ldots,-4m,2m+1,$$

and set $D_v = \alpha,\beta$.

(3) Suppose $v \equiv 7 \pmod{12}$ and let $v=12m+7$. Define sequences

$$\alpha = 4m+2,-(6m+3),4m+3,-(6m+2),\ldots,5m+2,-(5m+3), \text{ and}$$
$$\beta = 3m+2,-(3m+1),3m+3,-3m,\ldots,4m+1,-(2m+2),$$

and set $D_v = \alpha,\beta$. □

It remains an open question to determine the existence of Ucycles for TTS(v)s where $v \equiv 10 \pmod{12}$. For each $v \equiv 10 \pmod{12}$, there exists a 1-rotational TTS(v), but these designs also contain a short-orbit base block. While the same general methods are applicable, some adjustment must be made to the definition of difference sequence to accommodate short-orbit base blocks. Colbourn et al. have generated all TTS(10)s having no repeated blocks [4]. This listing was obtained by Dewar who then wrote a program to search for Ucycles of rank three for each of these designs. This search was not difficult to execute because once the first block and the order of appearance of points in this block has been fixed, the entry of new blocks into the sequence is forced. None of the simple TTS(10)s was found to admit a Ucycle of rank three for its blocks, and therefore no TTS(10) admits a Ucycle of rank three for its blocks. However, we conjecture that there exist higher order TTS(v)s with $v \equiv 10 \pmod{12}$ that do admit rank three Ucycles.

When $v \equiv 0,3,6,9 \pmod{12}$, it is impossible to create Ucycles of rank three for the blocks of a TTS(v) as three does not divide $v-1$ (see Lemma 3.5 on page 73); therefore, for TTS(v)s of these orders we investigate the existence of 2-intersecting cyclic Gray codes.

Lemma 5.7 ([9]). *Let v be odd and let $D_{v-1}=d_0,d_1,\ldots,d_{\ell-1}$ be a difference sequence of order $v-1$. If (1) d_0 is coprime to $v-1$, (2) $d_{\ell-1}=-(2n+1)d_0$, and (3) $d_{\ell-2}=2(2n+1)d_0$, for some $n \in \mathbb{Z}_{v-1}$ that is coprime to $(v-1)/2$, then there exists a 2-intersecting cyclic Gray code for the blocks of the 1-rotational TTS(v) induced by D_{v-1}.*

Proof. Suppose $D_{v-1}=d_0,d_1,\ldots,d_{\ell-1}$ is a difference sequence of order $v-1$ with d_0 coprime to $v-1$, $d_{\ell-2}=2(2n+1)d_0$ and $d_{\ell-1}=-(2n+1)d_0$, where $n \in \mathbb{Z}_{v-1}$

is coprime to $(v-1)/2$. Recall (from Lemma 5.5 (2)) that $U_{D_{v-1}} = \infty, 0, d_0, d_0 + d_1, d_0 + d_1 + d_2, \ldots, \sum_{i=0}^{\ell-1} d_i$, with addition performed modulo $v-1$, is the non-cyclic partial Ustring induced by D_{v-1}. Create $v-2$ other non-cyclic partial Ustrings by developing $U_{D_{v-1}}$ through $\mathbb{Z}_{v-1} \setminus \{0\}$. As $U_{D_{v-1}}$ represents the base blocks of a 1-rotational TTS(v), this collection of partial Ustrings represents all the blocks of the 1-rotational TTS(v). Let $L_{D_{v-1}}$ be the sequence of blocks represented by $U_{D_{v-1}}$:

$$\{\infty, 0, d_0\}, \{0, d_0, d_0 + d_1\}, \{d_0, d_0 + d_1, d_0 + d_1 + d_2\}, \ldots, \left\{\sum_{i=0}^{\ell-3} d_i, \sum_{i=0}^{\ell-2} d_i, \sum_{i=0}^{\ell-1} d_i\right\}.$$

Because we are constructing a cyclic Gray code, we will now work with $L_{D_{v-1},x}$, $x \in \{0, 1, \ldots, v-2\}$, where $L_{D_{v-1},x}$ denotes the sequence of blocks represented by the Ustring $U_{D_{v-1}} + x$. It remains to join these lists of blocks together to form a 2-intersecting cyclic Gray code; we must ensure that the blocks on either side of a join share two points.

The first block in $L_{D_{v-1}}$ is $\{\infty, 0, d_0\}$. This block shares two points with the first block of L_{D_{v-1},d_0} which is $\{\infty, d_0, 2d_0\}$. Write $L_{D_{v-1}}$ in reverse order and join it to L_{D_{v-1},d_0}. Write $L_{D_{v-1},2d_0}$ in reverse order and join it to $L_{D_{v-1},3d_0}$, and so on. As $v-1$ is even, the sequences $L_{D_{v-1},x}$, $x \in \{0, 1, \ldots, v-2\}$, can all be paired in this manner because d_0 is coprime to $v-1$. There are now $(v-1)/2$ sequences of the form $\overline{L_{D_{v-1},2md_0}}, L_{D_{v-1},(2m+1)d_0}$, $m \in \{0, 1, \ldots, (v-1)/2\}$, where $\overline{L_{D_{v-1}}}$ indicates that the sequence $L_{D_{v-1}}$ is written in reverse order. Each of these sequences starts with a block of the form $\{x, x + 2(2n+1)d_0, x + (2n+1)d_0\}$ and ends with a block of the form $\{x + d_0, x + 2(2n+1)d_0 + d_0, x + (2n+1)d_0 + d_0\}$, $x \in \mathbb{Z}_{v-1}$. We must now join these new sequences together such that the blocks on either side of each join share two points. Consider the block $B = \{x, x + 2(2n+1)d_0, x + (2n+1)d_0\}$ which appears at the beginning of the sequence $\overline{L_{D_{v-1}}}, L_{D_{v-1},d_0}$. The difference $(2n+1)d_0$ appears twice in B; therefore, $B + (2n+1)d_0$ has two points in common with B. The block $B + (2n+1)d_0 = \{x + (2n+1)d_0, x + 3(2n+1)d_0, x + 2(2n+1)d_0\}$ appears as the last block of $L_{D_{v-1},(2n+1)d_0}$. Since $\gcd(n, (v-1)/2) = 1$, we can join all sequences together to form a 2-intersecting cyclic Gray code as follows:

$$\overline{L_{D_{v-1}}}, L_{D_{v-1},d_0}, \overline{L_{D_{v-1},-2nd_0}}, L_{D_{v-1},-(2n-1)d_0}, \ldots, \overline{L_{D_{v-1},2nd_0}}, L_{D_{v-1},(2n+1)d_0}. \qquad \square$$

Consider the difference sequence $D_{14} = 3, -1, -6, 4, 5$ of order 14. The non-cyclic partial Ustring induced by this difference sequence is $U_{D_{14}} = \infty, 0, 3, 2, -4, 0, 5$ and this string represents the blocks $L_{D_{14}} = \{\infty, 0, 3\}$, $\{0, 3, 2\}$, $\{3, 2, 10\}$, $\{2, 10, 0\}$, $\{10, 0, 5\}$. Creating the intermediate sequences, we join $\overline{L_{D_{14}}}$ to $L_{D_{14},3}$ which yields $\{10, 0, 5\}$, $\{2, 10, 0\}$, $\{3, 2, 10\}$, $\{0, 3, 2\}$, $\{\infty, 0, 3\}$, $\{\infty, 3, 6\}$, $\{3, 6, 5\}$, $\{6, 5, 13\}$, $\{5, 13, 3\}$, $\{13, 3, 8\}$. We also join $\overline{L_{D_{14},6}}$ to $L_{D_{14},9}$ and $\overline{L_{D_{14},12}}$ to $L_{D_{14},1}$, and so on. Finally, we write $\overline{L_{D_{14}}}, L_{D_{14},3}, \overline{L_{D_{14},8}}, L_{D_{14},11}, \ldots, \overline{L_{D_{14},6}}, L_{D_{14},9}$. The complete 2-intersecting cyclic Gray code for the 1-rotational TTS(15) induced by D_{14} is given in Table 5.6.

Table 5.6 A 2-intersecting cyclic Gray code for the blocks of the 1-rotational TTS(15) with base blocks given by the non-cyclic partial Ustring $U_{D_{14}} = \infty, 0, 3, 2, -4, 0, 5$ (read column-wise)

$\{10,0,5\}$	$\{12,2,7\}$	$\{0,4,9\}$	$\{2,6,11\}$
$\{2,10,0\}$	$\{4,12,2\}$	$\{6,0,4\}$	$\{8,2,6\}$
$\{3,2,10\}$	$\{5,4,12\}$	$\{7,6,0\}$	$\{9,8,2\}$
$\{0,3,2\}$	$\{2,5,4\}$	$\{4,7,6\}$	$\{6,9,8\}$
$\{\infty,0,3\}$	$\{\infty,2,5\}$	$\{\infty,4,7\}$	$\{\infty,6,9\}$
$\{\infty,3,6\}$	$\{\infty,5,8\}$	$\{\infty,7,10\}$	$\{\infty,9,12\}$
$\{3,6,5\}$	$\{5,8,7\}$	$\{7,10,9\}$	$\{9,12,11\}$
$\{6,5,13\}$	$\{8,7,1\}$	$\{10,9,3\}$	$\{12,11,5\}$
$\{5,13,3\}$	$\{7,1,5\}$	$\{9,3,7\}$	$\{11,5,9\}$
$\{13,3,8\}$	$\{1,5,10\}$	$\{3,7,12\}$	$\{5,9,0\}$
$\{4,8,13\}$	$\{6,10,1\}$	$\{8,12,3\}$	
$\{10,4,8\}$	$\{12,6,10\}$	$\{0,8,12\}$	
$\{11,10,4\}$	$\{13,12,6\}$	$\{1,0,8\}$	
$\{8,11,10\}$	$\{10,13,12\}$	$\{12,1,0\}$	
$\{\infty,8,11\}$	$\{\infty,10,13\}$	$\{\infty,12,1\}$	
$\{\infty,11,0\}$	$\{\infty,13,2\}$	$\{\infty,1,4\}$	
$\{11,0,12\}$	$\{13,2,1\}$	$\{1,4,3\}$	
$\{0,12,7\}$	$\{2,1,9\}$	$\{4,3,11\}$	
$\{12,7,11\}$	$\{1,9,13\}$	$\{3,11,1\}$	
$\{7,11,2\}$	$\{9,13,4\}$	$\{11,1,6\}$	

Note that Lemma 5.7 may only be applied when $v \equiv 3,9 \pmod{12}$ as $v-1$ must be even.

Theorem 5.11 ([9]). *For each $v \equiv 3 \pmod{12}$, $v > 3$, there exists a TTS(v) that admits a 2-intersecting cyclic Gray code.*

Proof. For each $v \equiv 3 \pmod{12}$, $v > 3$, we prove that there exists a 1-rotational TTS(v) admitting a 2-intersecting cyclic Gray code. As we are working with 1-rotational designs, all operations will be done modulo $v-1$. The proof is broken into four cases: $v \equiv 3, 15, 27, 39 \pmod{48}$. For each case we give a difference sequence such that the properties required by Lemma 5.7 hold. We will give the proof in detail for the first case, but only the sequence in subsequent cases as the details are very similar to the first. In each case D_{v-1} consists of six subsequences. We will refer to these subsequences as α_1, α_2, β, γ_1, γ_2 and e, and we set $D_{v-1} = \alpha_1, \alpha_2, \beta, \gamma_2, \gamma_1, e$. The sixth subsequence, e, is a single element and in each case will represent itself. D_{v-1} is of length $v/3$. The length of α_1 is the same as that of γ_1 and the length of α_2 is the same as that of γ_2 (although these lengths differ for each case). We will see that the elements of γ_1 and γ_2 are related to those of α_1 and α_2 by the following rule. Let $\alpha_1 = x_0, x_1, \ldots, x_{\ell-2}$ and $\gamma_1 = y_0, y_1, \ldots, y_{\ell-2}$, then

$$y_{\ell-i} = f(x_i) = \begin{cases} (v-1)/2 - x_i & \text{if } x_i > 0 \\ -(v-1)/2 - x_i & \text{if } x_i < 0 \end{cases} \tag{5.1}$$

The same rule holds for α_2 and γ_2. As a result, σ_{γ_1} is the same as $\overline{\sigma}_{\alpha_1}$ with signs reversed. Similarly σ_{γ_2} is the same as $\overline{\sigma}_{\alpha_2}$ with signs reversed. In fact, equation (5.1) holds for the entire sequence D_{v-1} without e. Recall that the length of D_{v-1} must be $v/3$. While it may not be immediately obvious from the representations, it is not difficult to confirm that the sequence consisting of $\alpha_1, \alpha_2, \beta, \gamma_2, \gamma_1$ has the property that the sum of the ith element and the $((v/3-1)-i)$th element is either $-(v-1)/2$ or $(v-1)/2$ with each pair of summands of the same sign.

When $v \equiv 3,27 \pmod{48}$, the first entry of D_{v-1} is $(v+3)/6$. Due to the property stated above, when $v \equiv 3,27 \pmod{48}$, the second last element of D_{v-1} is $(v-3)/3$ (since $(v+3)/6+(v-3)/3=(v-1)/2$). Similarly, when $v \equiv 15,39 \pmod{48}$, the first entry of D_{v-1} is $-(v+3)/6$ and the second last element of D_{v-1} is $-(v-3)/3$. Lemma 5.7 requires that the first entry of D_{v-1} be coprime to $v-1$. To determine $\gcd((v+3)/6, v-1)$, substitute $v=12m+3$. Since $\gcd(2m+1, 12m+2)=\gcd(2m+1,-4)=1$, we conclude that $v-1$ is always coprime to $(v+3)/6$. Lemma 5.7 also requires that there exist an n such that $2(2n+1)d_0 = d_{v/3-2}$ and $-(2n+1)d_0 = d_{v/3-1}$, with all operations modulo $v-1$. It will be easy to confirm that these two equations hold once our sequences have been constructed.

To determine that an n with the above properties is coprime to $(v-1)/2$, notice that the summation property implies that $d_0 + d_{v/3-2} = d_0 + 2(2n+1)d_0 \equiv (v-1)/2 \pmod{(v-1)}$. Substituting $v=12m+3$ we obtain $(4n+3)d_0 \equiv (12m+2)/2 \pmod{v-1}$. We can divide this equality by 2 and remove d_0 as d_0 is coprime to $v-1$, which yields $2n \equiv 3m-1 \pmod{(v-1)/2}$. As $(v-1)/2 = 6m+1$, we have $(3m+1)2n \equiv (3m+1)(3m-1) \pmod{(6m+1)}$. Solving for n we obtain $n \equiv 9m^2 - 1 \pmod{(6m+1)}$. We will choose the minimal n that satisfies this equation. Finally, we must show that $\gcd(9m^2-1, 6m+1)=1$ in order to prove that n is coprime to $(v-1)/2$. It is clear that if $\gcd(2x,y)=1$, then $\gcd(x,y)=1$. Using this property, determine $\gcd(2(9m^2-1), 6m+1)$. This statement is equivalent to $\gcd(3m-1,3)$, which is clearly one. Therefore, n is coprime to $(v-1)/2$. For each case it remains to prove that the given D_{v-1} is a difference sequence.

Case 1: Suppose $v \equiv 3 \pmod{48}$ and let $v = 48m+3$. Define

$$\begin{aligned}
\alpha_1 &= 8m+1, -(10m+1), 10m+2, -10m, 10m+3, \ldots, \\
&\quad \ldots, -(9m+2), 11m+1, \\
\alpha_2 &= -15m, 13m-1, -(15m+1), 13m-2, \ldots, \\
&\quad \ldots, 12m+2, -(16m-2), 12m+1, -(16m-1), \\
\beta &= 8m-1, 20m+1, 8m-2, 20m, \ldots, 18m+2, 6m-1, \ldots, \\
&\quad \ldots, 4m+1, 16m+3, 4m, 16m+2, \\
\gamma_1 &= 13m, -(15m-1), 13m+1, -(15m-2), \ldots, \\
&\quad \ldots, 14m-2, -(14m+1), 14m-1, -14m, 16m, \\
\gamma_2 &= -(8m+2), 12m, -(8m+3), 12m-1, \ldots, \\
&\quad \ldots, 11m+3, -9m, 11m+2, -(9m+1).
\end{aligned}$$

Set $D_{v-1} = \alpha_1, \alpha_2, \beta, \gamma_2, \gamma_1, 16m+1$. Notice that the sequence α_1 follows a clear sequential pattern, except for the first entry. Equation (5.1) implies that γ_1 also follows a sequential pattern, except for its last entry. Ignoring signs, together the sequences α_1, α_2, β, γ_1 and γ_2 contain $\{4m, \ldots, 8m-1\} \cup \{8m+1, \ldots, 16m\} \cup \{16m+2, \ldots, 20m+1\} \subset \mathbb{Z}_{v-1}$. The sums of pairs of consecutive elements in α_1 are $\sigma_{\alpha_1} = -2m, 1, 2, \ldots, 2m-1$, and the sums of pairs of consecutive elements in α_2 are $\sigma_{\alpha_2} = -(2m+1), -(2m+2), \ldots, -(4m-3), -(4m-2)$. The sequence β has sums of consecutive pairs of elements $\sigma_\beta = 28m, 28m-1, \ldots, 24m+1, \ldots, 20m+3, 20m+2$, which is equivalent to $-(20m+2), -(20m+3), \ldots, 24m+1, \ldots, 20m+3, 20m+2$. Because the sums of consecutive pairs of elements in γ_1 are the same as those in α_1 with signs reversed (similarly for γ_2 and α_2), the sums of pairs of consecutive elements in the sequences α_1, α_2, β, γ_1 and γ_2 cover $\pm\{\{1, \ldots, 4m-2\} \cup \{20m+2, \ldots, 24m\}\} \cup \{24m+1\}$. The four sums of two consecutive elements appearing at the joins of sequences $\alpha_1, \alpha_2, \beta, \gamma_2$ and γ_1 are $\pm(4m-1)$ (between α_1 and α_2 and between γ_1 and γ_2) and $\pm 8m$ (between α_2 and β and between β and γ_2). Finally, $16m+1$ appears once as the last element in D_{v-1} and once in negative form as the sum across the join of γ_1 and $16m+1$ (because $32m+1 \equiv -(16m+1) \pmod{48m+2}$).

Case 2: Suppose $v \equiv 15 \pmod{48}$ and let $v = 48m+15$. A 2-intersecting cyclic Gray code for a 1-rotational TTS(15) is given in Table 5.6 (see page 133). For $v > 15$, define

$$\begin{aligned}
\alpha_1 &= -(8m+3), 12m+2, -(8m+4), 12m+1, \ldots, \\
&\quad \ldots, 11m+4, -(9m+2), \\
\alpha_2 &= 9m+3, -(11m+3), 9m+4, -(11m+2), \ldots, \\
&\quad \ldots, 10m+2, -(10m+4), 14m+4, -(12m+3), \\
\beta &= 4m+1, 16m+6, 4m+2, 16m+7, \ldots, 6m+1, 18m+6, \ldots, \\
&\quad \ldots, 8m, 20m+5, 8m+1, 20m+6, \\
\gamma_1 &= -(15m+5), 13m+3, \ldots, \\
&\quad \ldots, 12m+6, -(16m+3), 12m+5, -(16m+4), \\
\gamma_2 &= -(12m+4), 10m+3, -(14m+3), 14m+5, \ldots, \\
&\quad \ldots, -(13m+5), 15m+3, -(13m+4), 15m+4.
\end{aligned}$$

Set $D_{v-1} = \alpha_1, \alpha_2, \beta, \gamma_2, \gamma_1, -(16m+5)$.

Case 3: Suppose $v \equiv 27 \pmod{48}$ and let $v = 48m+27$. Define

$$\begin{aligned}
\alpha_1 &= 8m+5, -(10m+6), 10m+7, -(10m+5), 10m+8, \ldots \\
&\quad \ldots, 11m+6, -(9m+6),
\end{aligned}$$

$$\begin{aligned}
\alpha_2 &= 13m+6, -(15m+8), 13m+5, -(15m+9), \ldots \\
&\ldots, -(16m+5), 12m+8, -(16m+6), 12m+7, -(8m+6), \\
\beta &= 16m+10, 4m+2, 16m+11, 4m+3, \ldots, 6m+2, 18m+11, \ldots \\
&\ldots, 20m+10, 8m+2, 20m+11, 8m+3, \\
\gamma_1 &= -(15m+7), 13m+7, \ldots, \\
&\ldots, 14m+5, -(14m+8), 14m+6, -(14m+7), 16m+8, \\
\gamma_2 &= -(16m+7), 12m+6, -(8m+7), 12m+5, -(8m+8), \ldots \\
&\ldots, -(9m+5), 11m+7.
\end{aligned}$$

Set $D_{v-1} = \alpha_1, \alpha_2, \beta, \gamma_2, \gamma_1, 16m+9$.
Case 4: Suppose $v \equiv 39 \pmod{48}$ and let $v = 48m+39$. Define

$$\begin{aligned}
\alpha_1 &= -(8m+7), 10m+8, -(10m+9), 10m+7, -(10m+10), \ldots, \\
&\ldots, -(11m+8), 9m+8, \\
\alpha_2 &= -(13m+10), 15m+12, -(13m+9), 15m+13, \ldots, \\
&\ldots, 16m+11, -(12m+10), \\
\beta &= 20m+16, 8m+5, 20m+15, 8m+4, \ldots, 18m+15, 6m+4, \ldots, \\
&\ldots, 16m+15, 4m+4, 16m+14, 4m+3, \\
\gamma_1 &= 15m+11, -(13m+11), \ldots, \\
&\ldots, -(14m+9), 14m+12, -(14m+10), 14m+11, -(16m+12), \\
\gamma_2 &= -(12m+9), 8m+8, \ldots, \\
&\ldots, 9m+6, -(11m+10), 9m+7, -(11m+9).
\end{aligned}$$

Set $D_{v-1} = \alpha_1, \alpha_2, \beta, \gamma_2, \gamma_1, -(16m+13)$. □

The cases for which the existence of 2-intersecting Gray codes for TTS(v)s remain unknown are $v \equiv 0, 6, 9 \pmod{12}$. Dewar has constructed a 2-intersecting cyclic Gray code for a TTS(v) of each order $v \leq 200$ where $v \equiv 9 \pmod{12}$, but has not been able to determine a general construction. The cases $v \equiv 0, 6 \pmod{12}$ present special problems. For $v \equiv 0 \pmod{12}$, one must either deal with 1-rotational systems or cyclic systems containing short-orbit base blocks, while for $v \equiv 6 \pmod{12}$, one must deal with 1-rotational systems. It will be necessary to determine a new method for joining partial Ustrings together to create 2-intersecting Gray codes in these two cases.

Finally, we close this section noting that the results regarding the existence of 2-intersecting cyclic Gray codes (here we think of the rank three Ucycles as 2-intersecting cyclic Gray codes) for TTSs lead to the following two corollaries.

Corollary 5.1. *For each $v \equiv 1,3,4,7 \pmod{12}$, with exceptions as stated in Theorem 5.10, there exists a TTS(v) that admits an A_3-ordering (see Fig. 1.1 on page 4).*

Corollary 5.2. *For each $v \equiv 1,3,4,7 \pmod{12}$, with exceptions as stated in Theorem 5.10, there exists a TTS(v) that admits a $\{B_6, B_7\}$-ordering (see Fig. 4.10 on page 105).*

5.3 Gray Codes and Rank Two Universal Cycles for BIBDs and PBDs

We have seen that there exist infinite families of twofold triple systems that admit 2-intersecting cyclic Gray codes (see Sect. 5.2). What can be said about minimal change orderings for BIBD(v,k,λ)s in general? When index is one and strength is two, each pair of blocks in a BIBD(v,k,λ) intersect in at most one point. In this case, we can ask only for a weak minimal change property; consecutive blocks must share exactly one common element. The existence of such a minimal change ordering is equivalent to the existence of a Hamilton path or cycle in the 1 block-intersection graph. In Sect. 4.1.2 we saw that the existence of Hamilton cycles in 1 block-intersection graphs is entirely determined for PBD$(v,K,1)$s. Hare's result implies that if $\min(K) \geq 3$, a PBD$(v,K,1)$ admits a minimal change ordering for its blocks [13]. When the index is larger than one or the strength of the design is greater than two, stronger minimal change properties can be defined. Existence results in this area are scattered. M. Colbourn and Johnstone have discussed non-existence of 2-intersecting Gray codes for twofold triple systems [6], while Dewar has shown that 2-intersecting cyclic Gray codes exist for several infinite families of twofold triple systems [9]. Much work has been done regarding the existence of Gray codes and Ucycles for k-subsets of n-sets [3, 10, 11, 14, 26], and these results can be applied to designs of the form BIBD$\left(v,k,\binom{v-2}{k-2}\right)$. For $k=2$, Ucycles for ordered pairs have applications in statistical design. These are discussed in Sect. 6.5.2. Other Hamiltonicity results from Sect. 4.1.2 lead to orderings in which at most $k-1$ points change between consecutive blocks, but fewer may change. Thus, it is difficult to translate these results into the language of Gray codes and Ucycles. Furthermore, when $\lambda > 1$, consecutive blocks in a Hamilton cycle in the 1 block-intersection graph will share a single point; however, it is possible that an ordering in which consecutive blocks intersect in two (or more) points exists.

We turn now to the existence of Ucycles for BIBD(v,k,λ)s. Recall that Ucycles require an alphabet and a notion of rank of an object which corresponds to the size of subword which will represent the object. While block size is variable, the strength of all BIBD(v,k,λ)s is two; therefore, we investigate the existence of rank two Ucycles. While we are most interested in determining the existence of Ucycles for block designs, the existence of such an ordering has other implications. For example, a design admitting a Ucycle of rank two is an Eulerian design. A design is **Eulerian** if and only if the block-intersection graph of the design is Hamiltonian and there exists a Hamilton cycle in the block-intersection graph which does not pass through any 3-claws in the design. Note that this definition of Eulerian is not the same as one used in the context of hypergraphs [2].

As discussed in Sect. 3.2.3 and recalled above, when $\lambda = 1$, the best possible arrangement for the blocks of a BIBD(v,k,λ) is to have consecutive blocks intersecting in one element. In this case, a pair of points represents a unique block; therefore, a rank two Ucycle contains all the information necessary to reconstruct the design. However, when $\lambda > 1$, each pair of points appears in λ blocks of the design and so a pair of consecutive points in a rank two Ucycle does not uniquely determine a block of the BIBD. In order to recover the blocks of a BIBD(v,k,λ) with $\lambda > 1$ from a rank two Ucycle, we will need to provide some additional information.

In a Ucycle of rank two, each consecutive pair of points in the sequence represents a block. Furthermore, this pair of points must appear in the block in question. That is, if $\{x,y\}$ represents block B, then $\{x,y\} \in B$. This means that the alphabet of a rank two Ucycle for a BIBD, $S = (V,\mathscr{B})$, is $V' \subseteq V$. Note that if the blocks of a design are of size greater than two, there is more than one way to represent each block.

We begin with designs which do not admit rank two Ucycles. Then we move on to constructions of Ucycles of rank two; first for cyclic triple systems, then for cyclic designs. Both constructions are due to Dewar [9]. While she found these to be largely successful, other radically different approaches may be employed to show the existence (or non-existence) of Ucycles for BIBDs in general. Finally, we close this section with an instance where it is possible to construct rank two Ucycles for non-cyclic designs. As there are many questions remaining to be answered regarding Ucycles for BIBDs and PBDs, the reader is encouraged to peruse the Conjectures, and Exercises and Problems sections (see Sect. 5.3.3 and 5.3.3, beginning on page 168).

5.3.1 *Designs Which Cannot Admit Universal Cycles of Rank Two*

At this time, the existence of rank two Ucycles is known only for certain BIBD (v,k,λ)s; however, we suspect that almost all designs admit Ucycles of rank two. There are, however, some exceptions. A BIBD$(v,2,1)$ with v even does not admit a Ucycle of rank two. The PBD$(6,\{2,3\},1)$ consisting of the blocks of the Fano plane with one point removed does not admit a Ucycle of rank two. This design is not a member of the following infinite family of PBDs which also do not admit Ucycles of rank two.

Theorem 5.12 ([9]). *Given $v \in \mathbb{N}$ and any partition of $v = \sum_{i=1}^{\ell} v_i$ such that $v_1 \geq 3$ and $v_i \geq 2$, for all $i \in \{2,3,\ldots,\ell\}$, there exists a PBD$(v, 2 \cup \{v_i : 1 \leq i \leq \ell\}, 1)$ that does not admit a rank two Ucycle.*

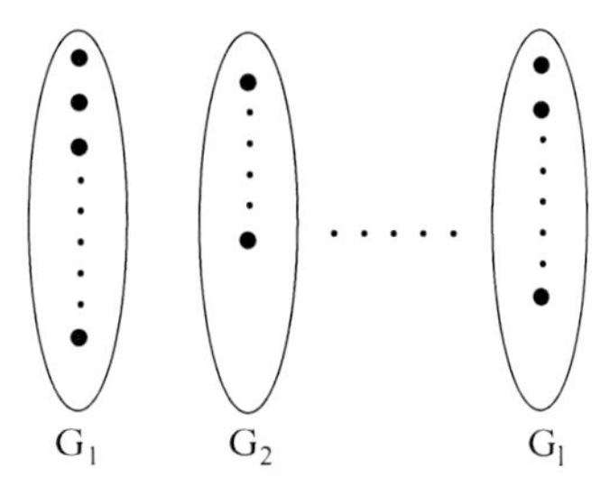

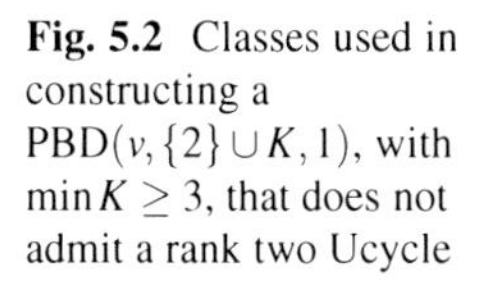
Fig. 5.2 Classes used in constructing a PBD$(v, \{2\} \cup K, 1)$, with $\min K \geq 3$, that does not admit a rank two Ucycle

Proof. Define a pairwise balanced design on v points as follows. Partition the points into ℓ disjoint classes $G_1, G_2, \dots, G_\ell$ (represented in Fig. 5.2) such that $|G_i| = v_i$, for all $i \in \{1, 2, \dots, \ell\}$. Let the blocks of the design be the classes of the partition and all pairs of points not in the same partition. That is, the design has point set $V = \cup_{i=1}^{\ell} G_i$ and blocks

$$\mathscr{B} = \{G_1, G_2, \dots, G_\ell\} \cup \{\{x, y\} : x, y \in V, x \in G_i, y \in G_j, i \neq j\}.$$

In a rank two Ucycle for this design, each class G_i will be represented by a pair of points from the class. The other blocks in the design are pairs of points and therefore will represent themselves. Each of these pairs must appear in a rank two Ucycle, thus, the existence of a rank two Ucycle is equivalent to the existence of an Euler tour in the graph having the points of $\cup_{i=1}^{\ell} G_i$ as vertices and edges the pairs representing each block. Denote this graph by H. For an Euler tour to exist, every vertex must have even degree. Consider a point p in G_1. The number of edges incident to p in H is $|G_2| + |G_3| + \dots + |G_\ell|$, unless p is one of the two points that represent G_1, in which case p has one additional edge incident. Since $|G_1| \geq 3$, G_1 contains at least one point of each type. Therefore, both types of vertex appear in H, implying that H cannot have an Euler tour. □

As blocks of size two appear to create problems in the construction of rank two Ucycles, in Sect. 5.3.3, we conjecture that they are the primary barrier to admitting Ucycles of rank two. At this time we can prove this conjecture only for certain cyclic designs; however, we see this as significant progress towards a general result.

5.3.2 *Universal Cycles of Rank Two for Cyclic BIBDs*

In this section, we focus on cyclic designs and prove that an infinite family of cyclic BIBDs admit Ucycles of rank two. Recall from Sect. 2.2.5 that, given a design $S = (V, \mathscr{B})$, the base block coloured pair adjacency graph, G_{base}^{V}, is a multigraph with vertex set V and an edge of colour B between vertices v_1 and v_2 if the difference $v_2 - v_1$ appears in the base block B. A Ucycle of rank two describes a tour in G_{base}^{V}. While each vertex of G_{base}^{V} may not be in this tour, this tour will include edge colour

B exactly $|\text{Dev}(B)|$ times. Given a Ucycle of rank two for a cyclic BIBD(v,k,λ) and the colour of the edge induced in G^V_{base} by each pair of consecutive elements in the Ucycle, we can completely recover the blocks of the design from the Ucycle. The edge colour is the minimal amount of additional information required to recover these blocks.

Throughout this section, we will be working with the points appearing in a given base block and the differences in that base block. Suppose $B_d = \{0, d_1, d_2, \ldots, d_{(k-1)}\}$ is a base block of a BIBD(v,k,λ). We will assume that the zero point appears in each base block and we will let $d_0 = 0$. The points in the base block B_d are d_i, $i \in \{0,1,\ldots,k-1\}$, while the differences in B_d are $\pm(d_i - d_j)$, $i,j \in \{0,1,\ldots,k-1\}$, $i \neq j$. Whenever $d_i - d_j$ is a difference in a base block, $d_j - d_i$ is also a difference in the same base block, thus, we often talk about the differences $d_i - d_j$ with $i > j$. That is, we are interested in the $\binom{k}{2}$ unordered pairs of points in the base block. Any block in Dev(B_d) can be chosen as the base block used to generate this set of blocks; however, many blocks in Dev(B_d) do not contain the point zero and hence are not suitable as base blocks under our assumption that base blocks contain the point zero. We will often wish to work with a representative of Dev(B_d) that has a given difference also appearing as one of the points in the base block. In this case, simply shift the base block B_d, by adding a fixed value to every entry in B_d, to obtain a base block containing the desired difference as a point. For example, if d_t is a difference in $B_d = \{0, d_1, \ldots, d_{(k-1)}\}$ with $d_t = d_i - d_j$, then d_t appears (as a point) in $-d_j + B_d$ (note that zero is also a point in this new expression of the base block).

While a Ucycle of rank two is a sequence of points in which consecutive pairs represent blocks, it may help to visualize a Ucycle as a cycle in the graph theoretic sense. That is, each point of the Ucycle is a vertex and there is an edge between each consecutive pair of points representing the block induced by the endpoints. To save space we will often write a Ucycle as a sequence of points; however, to illustrate some concepts, we will write a Ucycle as a cycle as described. When edges of this cycle are coloured to represent the base block from which the block represented is developed, this is a subgraph of the base block coloured pair adjacency graph for the design. In almost all cases, taking a consecutive pair of points in a Ucycle and knowing the colour of the edge connecting these two points uniquely determines the block represented. However, it is possible that two different blocks developed from the same base block are represented by the same pair. In this case we simply designate one occurrence of the pair to represent one block and the other occurrence of the pair to represent the other block. This ensures that every block of the design appears exactly once in the rank two Ucycle for the design.

Throughout this section, given a cyclic BIBD(v,k,λ), let n_1 denote the number of full-orbit base blocks in the BIBD, let n_2 denote the number of non-regular short-orbit base blocks in the BIBD, and let n_3 denote the number of regular short-orbit base blocks in the BIBD. Let $\phi(x)$, $x \in \mathbb{Z}^+$, denote the Euler totient function which counts the number of positive integers less than or equal to x that are coprime to x.

Theorem 5.13 ([9]). *Let S be a cyclic BIBD(v,k,λ) with $k \geq 3$. If*

$$n_2 \leq \left\lceil \frac{\lambda\phi(v)}{k(k-1)} \right\rceil - 4 \quad \textit{and} \quad n_1 + n_2 \geq \left\lfloor \frac{n_3}{2} \right\rfloor + 3,$$

then S admits a Ucycle of rank two.

Before proving Theorem 5.13, it is necessary to establish several auxiliary results. These results will provide the tools for a constructive proof of the existence of Ucycles of rank two for cyclic BIBD(v,k,λ)s having the properties specified in Theorem 5.13. The process of determining a rank two Ucycle for a cyclic BIBD(v,k,λ) begins by concentrating on a single base block. Given a base block and the blocks developed from it, we determine a rank two Ucycle for these blocks. A cyclic sequence of points that has the properties of a rank two Ucycle and represents some, but not all, blocks of a design will be called a **partial Ucycle** of rank two. There are some base blocks for which it is not possible to construct a rank two partial Ucycle. We deal with these blocks later when we join partial Ucycles representing the blocks developed from individual base blocks together to form a rank two Ucycle for all blocks of the design. Throughout the construction process, we will be careful to avoid creating an ordering of blocks which cannot be represented as a rank two Ucycle. In particular, we wish to avoid creating an ordering in which three consecutive blocks intersect in exactly one point and where this is the only intersection point for each of the two consecutive pairs of blocks from the triple. This is equivalent to saying we must avoid the claw configuration on three consecutive blocks. For triple systems, the 3-claw is the configuration denoted B_3 (see Fig. 1.2 on page 5). For blocks of size k, the 3-claw is the $(3k-2,3)$-configuration in which one point appears in all three lines (as distinct from the other $(3k-2,3)$-configuration which is the **3**-path). An ordering which avoids the 3-claw will be said to have **Property *Π***. For the remainder of this section we will assume the Ucycles referred to are of rank two.

Lemma 5.8 ([9]). *Let S be a cyclic BIBD(v,k,λ), $k \geq 3$, and let $B_d = \{0, d_1, \ldots, d_{(k-1)}\}$ be a full-orbit base block of S. If* $\gcd(d_1,v) = 1$, *then there exists a partial Ucycle containing the blocks of* Dev(B_d). *We call this a **length-v partial Ucycle**.*

Proof. Let S be a cyclic BIBD(v,k,λ), $k \geq 3$, let $B_d = \{0, d_1, \ldots, d_{(k-1)}\}$ be a full-orbit base block of S and suppose d_1 is coprime to v. The cyclic sequence $U(B_d,d_1) = 0, d_1, 2d_1, \ldots, (v-1)d_1$, with multiplication performed modulo v, contains v unique elements and thus v unique consecutive pairs of points. Each consecutive pair of points represents a distinct block of Dev(B_d): $\{xd_1, (x+1)d_1\}$ represents $xd_1 + B_d$, $x \in \mathbb{Z}_v$. Furthermore, the blocks represented by consecutive pairs of points have property Π. To see this, consider four consecutive points in $U(B_d,d_1)$: $(a-1)d_1, ad_1, (a+1)d_1, (a+2)d_1$, where $a \in \mathbb{Z}_v$. These points represent the blocks $(a-1)d_1 + B_d$, $ad_1 + B_d$, and $(a+1)d_1 + B_d$. Written out in full, these blocks are

Table 5.7 The blocks of Dev($\{0,1,4\}$) in the order given by the length-15 partial Ucycle $U = 0,1,2,3,4,5,6,7,8,9,10,11,12,13,14$ (read column-wise)

$\{0,1,4\}$	$\{5,6,9\}$	$\{10,11,14\}$
$\{1,2,5\}$	$\{6,7,10\}$	$\{11,12,0\}$
$\{2,3,6\}$	$\{7,8,11\}$	$\{12,13,1\}$
$\{3,4,7\}$	$\{8,9,12\}$	$\{13,14,2\}$
$\{4,5,8\}$	$\{9,10,13\}$	$\{14,0,3\}$

$$\{(a-1)d_1, ad_1, (a-1)d_1+d_2, \ldots, (a-1)d_1+d_{(k-1)}\},$$
$$\{ad_1, (a+1)d_1, ad_1+d_2, \ldots, ad_1+d_{(k-1)}\}, \text{ and}$$
$$\{(a+1)d_1, (a+2)d_1, (a+1)d_1+d_2, \ldots, (a+1)d_1+d_{(k-1)}\}.$$

It is clear that the first two blocks intersect in ad_1 and the second two intersect in $(a+1)d_1$. These two (consecutive) intersection points are not the same because $d_1 \neq 0$. □

Notice that consecutive blocks represented by the length-v partial Ucycle constructed in the proof of Lemma 5.8 may intersect in more than one point. While the number of points in the intersection of two consecutive blocks is forced when we look at Ucycles of rank k for BIBD(v,k,λ)s, in Ucycles of rank two we require only that each pair of consecutive blocks intersect in at least one point.

Suppose we are looking at a cyclic STS(15) generated by the base blocks $\{0,5,10\}$, $\{0,1,4\}$, and $\{0,2,8\}$. Note that the base block $\{0,5,10\}$ is a regular short-orbit base block and contains no difference coprime to v; therefore, Lemma 5.8 cannot be applied to it. However, the base block $\{0,1,4\}$ contains the difference 1 which is clearly coprime to 15. The cycle $U = 0,1,2,3,4,5,6,7,8,9,10,11,$ $12,13,14$ is a length-15 partial Ucycle representing the ordering of Dev($\{0,1,4\}$) given in Table 5.7. Note that the cycle $U = 0,4,8,12,1,5,9,13,2,6,10,14,3,7,11$ is also a Ucycle for this set of blocks (though not the same ordering of the blocks).

Lemma 5.8 can be applied using any of the differences appearing in B_d as long as the difference is coprime to v. If B_d is a full-orbit base block and no difference in B_d is coprime to v, we apply the following lemma.

Lemma 5.9 ([9]). *Let S be a cyclic BIBD(v,k,λ), $k \geq 3$, and let $B_d = \{0, d_1, \ldots, d_{(k-1)}\}$ be a full-orbit base block of S. If* $\gcd(d_1, v) > 1$, *then there exists a collection of partial Ucycles (at least two) which considered together include each of the blocks of* Dev(B_d) *exactly once. If the partial Ucycles are each of length ℓ (they will all be the same length), there will be v/ℓ partial Ucycles in the collection. We call this a* ***collection of length-ℓ partial Ucycles****.*

The proof of Lemma 5.9 is similar to that of Lemma 5.8 and is thus left to the reader.

Suppose we are looking at a TTS(24) that has base block $\{0,3,12\}$, amongst others. The differences in this base block are $3,9,12,12,15,21$; therefore, each difference has a greatest common divisor with 24 of three or greater. Select 3 as the difference with which to create the collection of partial Ucycles representing

Table 5.8 The blocks of Dev($\{0,3,12\}$) in the order given by the collection of length-8 partial Ucycles $U_0 = 0,3,6,9,12,15,18,21$, $U_1 = 1,4,7,10,13,16,19,22$, and $U_2 = 2,5,8,11,14,17,20,23$

U_0	U_1	U_2
$\{0,3,12\}$	$\{1,4,13\}$	$\{2,5,14\}$
$\{3,6,15\}$	$\{4,7,16\}$	$\{5,8,17\}$
$\{6,9,18\}$	$\{7,10,19\}$	$\{8,11,20\}$
$\{9,12,21\}$	$\{10,13,22\}$	$\{11,14,23\}$
$\{12,15,0\}$	$\{13,16,1\}$	$\{14,17,2\}$
$\{15,18,3\}$	$\{16,19,4\}$	$\{17,20,5\}$
$\{18,21,6\}$	$\{19,22,7\}$	$\{20,23,8\}$
$\{21,0,9\}$	$\{22,1,10\}$	$\{23,2,11\}$

Dev($\{0,3,12\}$). As $\text{lcm}(3,24) = 24$, the blocks of Dev($\{0,3,12\}$) can be represented by three partial Ucycles, each of length eight. The partial Ucycles

$$U_0 = 0,3,6,9,12,15,18,21$$

$$U_1 = 1,4,7,10,13,16,19,22$$

$$U_2 = 2,5,8,11,14,17,20,23$$

represent the ordering of blocks given in Table 5.8.

Note that when a base block B_d contains the difference $v/2$ and this difference is used to generate the collection of partial Ucycles representing the blocks of Dev(B_d), the collection will consist of $v/2$ length-2 partial Ucycles: $U_0 = 0, v/2$, $U_1 = 1, v/2+1, \ldots, U_{v/2-1} = v/2-1, v-1$. Note that each of these sequences represents two blocks. For example, if $B_d = \{0, v/4, v/2\}$, the sequence $U_0 = 0, v/2$ represents the blocks $\{0, v/4, v/2\}$ and $\{v/2, 0, 3v/4\}$.

In practice, given a full-orbit base block, we will not know which difference was used to create the partial Ucycle (or Ucycles) representing the set of blocks generated by this base block; therefore, we refer to the partial Ucycle(s) representing the blocks developed from a single full-orbit base block (i.e. generated using Lemmas 5.8 or 5.9) as a collection of length-ℓ partial Ucycles, where ℓ is between 2 and v.

Before delving further into the construction of Ucycles for cyclic BIBDs in general, we consider the existence of Ucycles for symmetric designs.

Corollary 5.3. *Every cyclic symmetric BIBD*(v,k,λ)*,* $k \geq 3$*, admits a Ucycle of rank two.*

Proof. A cyclic symmetric design is generated by exactly one full-orbit base block. As each difference in $\mathbb{Z}_v$ appears λ times in this base block, Lemma 5.8 can be applied using the difference 1. The full-length partial Ucycle for the blocks developed from this base block is a Ucycle of rank two for the design. □

Lemmas 5.8 and 5.9 imply that a collection of partial Ucycles containing all blocks developed from a given base block can be created as long as the base block is not a short-orbit base block. In fact, a collection of partial Ucycles containing all blocks developed from a given base block can be created as long as the base block is not a *regular* short-orbit base block. We prove this statement in Lemmas 5.10 and 5.11; however, we first discuss the forms of short-orbit base blocks. A regular short-orbit base block of a cyclic BIBD(v,k,λ) has the form $\{0, v/k, 2v/k, \ldots, (k-1)v/k\}$. Such a base block generates v/k blocks which are pairwise disjoint. It is not possible to create a partial Ucycle of the form described in Lemmas 5.8 and 5.9 to represent the blocks developed from a regular short-orbit base block unless the base block appears k times in the design. In this unusual case, the sequence $U_0 = 0, v/k, 2v/k, \ldots, (k-1)v/k$ is a partial Ucycle representing k copies of the block $\{0, v/k, 2v/k, \ldots, (k-1)v/k\}$. The collection of length-(v/k) partial Ucycles formed by adding i to the points of U_0, for each $i \in \{0, 1, \ldots, (v-k)/k\}$, represents the blocks generated by k copies of the regular short-orbit base block. In general, this situation will not occur and the blocks developed from regular short-orbit base blocks will need to be inserted into existing collections of partial Ucycles. This issue will be addressed in Lemmas 5.13 and 5.14.

Short-orbit base blocks, whether regular or non-regular, have the following properties. The number of blocks generated by a short-orbit base block is a divisor of v. The number of blocks generated by a short-orbit base block (containing the zero point) must appear as a point in the base block. That is, if B_d is a short-orbit base block with $0 \in B_d$ and $n = |\mathrm{Dev}(B_d)|$, then $n \in B_d$. Furthermore, since $n + B_d = B_d$, the point $2n$ must be in B_d. Continuing this argument implies that B_d must contain the set of points $\{0, n, 2n, \ldots, (x-1)n\}$, where x is the minimal value in $\mathbb{Z}_v$ such that $xn \equiv 0 \pmod{v}$. Non-regular short-orbit base blocks must contain at least one difference that is not a multiple of n; therefore, each non-regular short-orbit base block must contain at least the following set of points $\{0, n, 2n, \ldots, (x-1)n, d, d+n, d+2n, \ldots, d+(x-1)n\}$, $d \in \mathbb{Z}_v$, where $n \nmid d$. A base block containing exactly these points generates the fewest blocks of any non-regular short-orbit base block. In this case $x = k/2$ and since n is the smallest number such that $xn \equiv 0 \pmod{v}$, $n = 2v/k$. Hence, each non-regular short-orbit base block generates at least $2v/k$ blocks. Note that every difference appearing in a short-orbit base block appears at least twice.

To better understand short-orbit base blocks, we present the possible short-orbit base blocks for cyclic BIBD(v,k,λ)s with small k. The only short-orbit base block possible in a cyclic triple system is the regular short-orbit base block $\{0, v/3, 2v/3\}$. Regular short-orbit base blocks can only appear in TS(v,λ)s with $v \equiv 3 \pmod{6}$. The short-orbit base blocks associated with cyclic BIBD$(v,4,\lambda)$s have one general form. The regular short-orbit base block $\{0, v/4, v/2, 3v/4\}$ is a special case of the short-orbit base block $\{0, v/2, u, v/2+u\}, 0 < u < v/2$. The only short-orbit base block possible in a cyclic BIBD$(v,5,\lambda)$ is the regular one. Finally, cyclic BIBD$(v,6,\lambda)$s may contain short-orbit base blocks of the following two forms: $\{0, v/2, u, u+v/2, w, w+v/2\}$, $u \neq w$, $0 < u, w < v/2$ and $\{0, v/3, 2v/3, u, u+v/3, u+2v/3\}, 0 < u \leq v/2$ (with the regular short-orbit base block $\{0, v/6, 2v/6, 3v/6, 4v/6, 5v/6\}$ a special case of each form).

The structure of non-regular short-orbit base blocks means that at least one point in the base block is not a multiple of the number of blocks generated by the base block which yields the following lemma. The reader is invited to complete the proof.

Table 5.9 The blocks of Dev($\{0,3,10,13\}$) in the order given by the length-10 partial Ucycle $U = 0,3,6,9,2,5,8,1,4,7$ (read column-wise)

$\{0,3,10,13\}$	$\{5,8,15,18\}$
$\{3,6,13,16\}$	$\{8,11,18,1\}$
$\{6,9,16,19\}$	$\{1,4,11,14\}$
$\{9,12,19,2\}$	$\{4,7,14,17\}$
$\{2,5,12,15\}$	$\{7,10,17,0\}$

Table 5.10 The blocks of Dev($\{0,10,20,2,12,22\}$) in the order given by the collection of length-5 partial Ucycles $U_0 = 0,2,4,6,8$ and $U_1 = 1,3,5,7,9$

U_0	U_1
$\{0,10,20,2,12,22\}$	$\{1,11,21,3,13,23\}$
$\{2,12,22,4,14,24\}$	$\{3,13,23,5,15,25\}$
$\{4,14,24,6,16,26\}$	$\{5,15,25,7,17,27\}$
$\{6,16,26,8,18,28\}$	$\{7,17,27,9,19,29\}$
$\{8,18,28,10,20,0\}$	$\{9,19,29,11,21,1\}$

Lemma 5.10 ([9]). *Let S be a cyclic BIBD(v,k,λ), $k \geq 3$, and let $B_d = \{0,d_1,\ldots,d_{(k-1)}\}$ be a non-regular short-orbit base block of S. Let d_1 be a point that is not a multiple of $|\mathrm{Dev}(B_d)|$. If $\gcd(d_1,|\mathrm{Dev}(B_d)|) = 1$, then there exists a partial Ucycle containing the blocks of* Dev(B_d).

Suppose $\{0,3,10,13\}$ is a base block of a BIBD$(20,4,2)$. This short-orbit base block generates ten blocks and $\gcd(3,10) = 1$. The cycle $U = 0,3,6,9,2,5,8,1,4,7$ is a length-10 partial Ucycle representing the blocks of Dev($\{0,3,10,13\}$) in the order given in Table 5.9. Note that each of the first three blocks is represented by the first two points in the block, but, the block $\{9,12,19,2\}$ is represented by $\{9,2\}$, where the 2 is obtained by adding $3+3+3=9$ to 13, which is the last point in the first block. This change in the position (within a block) of the representative points also occurs for blocks $\{8,11,18,1\}$ and $\{7,10,17,0\}$.

If the point, d_1, chosen for development of a partial Ucycle for the blocks of B_d is not coprime to $|\mathrm{Dev}(B_d)|$, then the following lemma applies.

Lemma 5.11 ([9]). *Let S be a cyclic BIBD(v,k,λ), $k \geq 3$, and let $B_d = \{0,d_1,\ldots\ldots,d_{(k-1)}\}$ be a non-regular short-orbit base block of S. Let d_1 be a point that is not a multiple of $|\mathrm{Dev}(B_d)|$. If $\gcd(d_1,|\mathrm{Dev}(B_d)|) > 1$, then there exists a collection of partial Ucycles (at least two) which considered together include each of the blocks of* Dev(B_d) *exactly once. If the partial Ucycles are each of length ℓ (they will all be the same length) there will be $|\mathrm{Dev}(B_d)|/\ell$ partial Ucycles in the collection.*

Again, the proof is a variant of Lemma 5.8 and the reader is invited to complete it.

Suppose $\{0,10,20,2,12,22\}$ is a base block of a BIBD$(30,6,2)$. This short-orbit base block generates ten blocks and no point in the base block is coprime to ten. Select 2 as the difference with which to create the collection of partial Ucycles. As $\mathrm{lcm}(2,10) = 10$, the blocks of Dev($\{0,10,20,2,12,22\}$) can be represented by two partial Ucycles, each of length five. The partial Ucycles $U_0 = 0,2,4,6,8$ and $U_1 = 1,3,5,7,9$ represent the ordering of blocks given in Table 5.10.

As the blocks developed from fewer than k copies of a regular short-orbit base block cannot be represented in a partial Ucycle created by developing one of the points in the base block, we turn to inserting these blocks into existing collections of partial Ucycles. Some insertion methods require a collection of partial Ucycles representing the blocks developed from a full-orbit base block, while others can be executed in collections of partial Ucycles representing blocks developed from a full or a non-regular short-orbit base block. If we need not distinguish between full orbit and non-regular short-orbit base blocks, we group these two types of base block under the heading **non-regular base blocks**. The lower bounds on the number of full orbit and non-regular short-orbit base blocks in the statement of Theorem 5.13 are required to ensure that there will always be enough partial Ucycles to allow for the insertion of all regular short-orbit blocks. When inserting into a collection of partial Ucycles representing the blocks of a full-orbit base block, $B_d = \{0, d_1, \ldots, d_{(k-1)}\}$, it is useful to keep the following rule in mind. Let $g = \gcd(d_1, v)$ and suppose d_1 was used to form a collection of length-ℓ partial Ucycles, $U(B_d, d_1)$, representing the blocks developed from B_d. If $g = 1$, then $\ell = v$ and $U(B_d, d_1)$ is a single full-length partial Ucycle. If $g > 1$, then $U(B_d, d_1)$ is a collection of length-ℓ partial Ucycles, where $\ell \leq v/2$. The number of partial Ucycles in $U(B_d, d_1)$ is g.

Blocks developed from a regular short-orbit base block will always be inserted into partial Ucycles so that consecutive blocks in the new sequence intersect. Thus, when adding points to a partial Ucycle or when joining partial Ucycles together, it remains to avoid creating the claw configuration on three consecutive blocks. The formulation of a Ucycle as a list of points from V allows us to avoid claw creation altogether (i.e. we will always maintain property Π). A claw involving the point x would be represented by the sequence y, x, x, z, where $x \neq y, z$. However, as each consecutive pair of points in the sequence represents a block containing this pair of points, this sequence implies that some block contains the point x twice—impossible. We now present several lemmas dealing with how and when we can insert regular short-orbit blocks, but first we ensure the existence of certain differences in full-orbit base blocks.

Lemma 5.12 ([9]). *Let S be a cyclic BIBD(v,k,λ), $k \geq 3$. Every full-orbit base block in S contains a difference that is not a multiple of v/k modulo v.*

Proving this lemma is Exercise 5.6.

Lemma 5.13 ([9]). *Let S be a cyclic BIBD(v,k,λ), $k \geq 3$, with a regular short-orbit base block B_e and a full-orbit base block $B_d = \{0, d_1, \ldots, d_{(k-1)}\}$. Select a difference $d_s \in B_d$ that is not a multiple of v/k modulo v. Let $U(B_d, d_1) = \{U_0, U_1, \ldots, U_{m-1}\}$, $1 \leq m \leq v/2$, be a collection of length-ℓ partial Ucycles representing the blocks of* Dev(B_d)*, with $d_1 \neq d_s$ used to form the partial Ucycles. Two distinct blocks of* Dev(B_e) *can be inserted into the partial Ucycles of $U(B_d, d_1)$.*

Proof. Let S be a cyclic BIBD(v,k,λ), $k \geq 3$, and let $B_d = \{0, d_1, \ldots, d_{(k-1)}\}$ be a full-orbit base block of the design. Select a difference d_s from B_d that is not a multiple of v/k modulo v; Lemma 5.12 implies that such a difference must exist. Let $U(B_d, d_1) = \{U_0, U_1, \ldots, U_{m-1}\}$ be a collection of length-ℓ partial Ucycles representing the blocks of Dev(B_d) with $d_1 \neq d_s$ used to form the partial

Ucycles. We will insert two blocks of $\mathrm{Dev}(B_e)$ into the partial Ucycles of $U(B_d, d_1)$. The blocks we insert are of the form $x + \{0, v/k, \ldots, (k-1)v/k\}$ and $(x + d_s) + \{0, v/k, \ldots, (k-1)v/k\}$, for some $x \in \{0, 1, \ldots, v/k - 1\}$. These two blocks are distinct because the only way to obtain a copy of $x + B_e$ is to add a multiple of v/k to $x + B_e$.

Select two points $y, z \in x + \{0, v/k, \ldots, (k-1)v/k\}$, $y \neq z$, to represent the block $x + B_e$. Then $(x + d_s) + \{0, v/k, \ldots, (k-1)v/k\}$ will be represented by the pair $\{y + d_s, z + d_s\}$. As each point of $\mathbb{Z}_v$ appears in a partial Ucycle of $U(B_d, d_1)$ exactly once (refer to Lemmas 5.8 and 5.9), there exists $i \in \{0, \ldots, m-1\}$ such that $y \in U_i$ and there exists $j \in \{0, \ldots, m-1\}$ such that $z \in U_j$. The method used to insert these two blocks depends on the relationship of i to j.

Case 1: $i = j$. In this case y and z appear in the same partial Ucycle of $U(B_d, d_1)$. Represent y and z in terms of d_1: $y = ad_1 + i$ and $z = cd_1 + i$, for $a, c \in \mathbb{Z}_v$. Without loss of generality, assume $a < c$. To insert the blocks represented by $\{y, z\}$ and $\{y + d_s, z + d_s\}$ into U_i, break U_i between the points $ad_1 + i$ and $(a+1)d_1 + i$ and between the points $(c-1)d_1 + i$ and $cd_1 + i$. We now have two segments: $(a+1)d_1 + i, (a+2)d_1 + i, \ldots, (c-1)d_1 + i$ and $cd_1 + i, (c+1)d_1 + i, \ldots, ad_1 + i$. Add $ad_1 + i$ to the beginning of the first segment and $cd_1 + i$ to the end of the first segment, yielding the segment $ad_1 + i, (a+1)d_1 + i, (a+2)d_1 + i, \ldots, (c-1)d_1 + i, cd_1 + i$. Reverse the orientation of this segment and reattach it to the segment running from $cd_1 + i$ to $ad_1 + i$. Now the partial Ucycle has two pairs of consecutive points $\{ad_1 + i, cd_1 + i\}$; therefore, change the $ad_1 + i$ adjacent to $(a+1)d_1 + i$ to $ad_1 + d_s + i$ and change the $cd_1 + i$ adjacent to $(c+1)d_1 + i$ to $cd_1 + d_s + i$. We have added two new blocks to the partial Ucycle; the first is represented by the pair $\{ad_1 + i, cd_1 + i\} = \{y, z\}$ and the second by $\{ad_1 + d_s + i, cd_1 + d_s + i\} = \{y + d_s, z + d_s\}$. All of the blocks represented by the original partial Ucycle U_i remain intact; however, the block $i + \{ad_1, (a+1)d_1, ad_1 + d_2, \ldots, ad_1 + d_s, \ldots, ad_1 + d_{(k-1)}\}$ which was originally represented by the pair $\{ad_1 + i, (a+1)d_1 + i\}$ is now represented by $\{ad_1 + d_s + i, (a+1)d_1 + i\}$ and the block $i + \{cd_1, (c+1)d_1, cd_1 + d_2, \ldots, cd_1 + d_s, \ldots, cd_1 + d_{(k-1)}\}$ which was originally represented by the pair $\{cd_1 + i, (c+1)d_1 + i\}$ is now represented by $\{cd_1 + d_s + i, (c+1)d_1 + i\}$. We will refer to this type of insertion as **Insertion *A***.

The procedure is better illustrated with a picture. In Fig. 5.3, green lines and numbers represent changes to the original partial Ucycle—the green lines are new connections and the green numbers are altered numbers (with original numbers in square brackets). The gray letters and numbers represent the segment of the original partial Ucycle that has been reversed (and which appears in its new orientation at the top of the figure).

Note that our choice of the segment $ad_1 + i, (a+1)d_1 + i, \ldots, (c-1)d_1 + i, cd_1 + i$ for reversal is arbitrary. We could just as easily reverse the segment $cd_1 + i, (c+1)d_1 + i, \ldots, (a-1)d_1 + i, ad_1 + i$. As above, we would then change the $ad_1 + i$ adjacent to $(a+1)d_1 + i$ to $ad_1 + d_s + i$ and change the $cd_1 + i$ adjacent to $(c+1)d_1 + i$ to $cd_1 + d_s + i$.

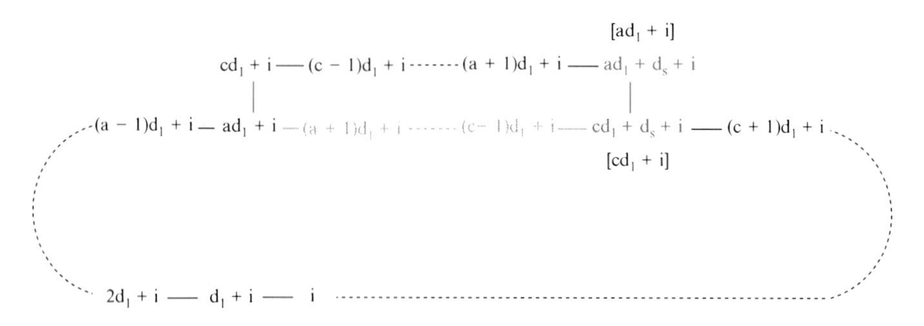

Fig. 5.3 Insertion A: two distinct blocks developed from a regular short-orbit base block are inserted into a single partial Ucycle

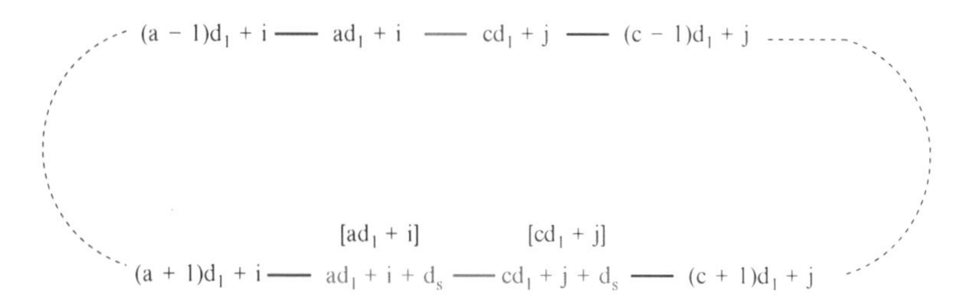

Fig. 5.4 Insertion B: two distinct blocks developed from a regular short-orbit base block are inserted between two partial Ucycles from the same collection

Case 2: $i \neq j$. This insertion occurs between two partial Ucycles (i.e. y and z appear in different partial Ucycles). The insertion is illustrated in Fig. 5.4 but we leave the proof of validity of this insertion to the reader. We will refer to this type of insertion as **Insertion *B***. In Fig. 5.4, green lines and numbers represent changes to the original Ucycles—the green lines are new connections and the green numbers are altered numbers (with original numbers in square brackets). The blue numbers and lines represent U_i (the partial Ucycle containing y) and the red letters and numbers represent U_j (the partial Ucycle containing z). □

We note that the method for inserting blocks into partial Ucycles described in the proof of Lemma 5.13 is reminiscent of that employed by Vickers and Silverman in creating balanced Gray codes (see page 35) [23].

It is possible to exactly determine which partial Ucycles the points y and z (defined in the proof of Lemma 5.13) appear on. Without loss of generality, assume the pair of points representing one of the insertion blocks is y and $z = y + v/k$. Suppose we have the collection of partial Ucycles $U(B_d, d_1) = \{U_0, U_1, \ldots, U_{m-1}\}$ representing the blocks of $\mathrm{Dev}(B_d)$, where $B_d = \{0, d_1, d_2, \ldots, d_{(k-1)}\}$. Further, suppose $d_s \in B_d$ with $d_s \neq d_1$ and $d_s \neq r \cdot v/k$, for all $r \in \{0, 1, \ldots, (k-1)\}$. As each point of $\mathbb{Z}_v$ appears in a partial Ucycle of $U(B_d, d_1)$ exactly once, there exists $i \in \{0, \ldots, m-1\}$ such that $y \in U_i$. This implies that $y = ad_1 + i$, for some $a \in \mathbb{Z}_v$.

On which sequence of $U(B_d,d_1)$ does $y+v/k$ appear? The answer depends on the divisibility of v by kg, where $g=\gcd(d_1,v)$ (recall that d_1 is the difference used to create the collection of partial Ucycles $U(B_d,d_1)$). If $kg|v$, then $y+v/k$ appears in U_i. The expression $kg|v$ implies $g|(v/k)$, thus $qg=v/k$, for some $q\in\mathbb{Z}$. In group theoretic terms, v/k belongs to the cyclic subgroup of $\mathbb{Z}_v$ generated by g, that is $v/k\in\langle g\rangle$. Now $d_1=sg$, for some $s\in\mathbb{Z}$, so $d_1\in\langle g\rangle$. Any element coprime to the order of the cyclic subgroup can be used to generate the subgroup; therefore, $\langle g\rangle=\langle d_1\rangle$ and $v/k\in\langle d_1\rangle$, which implies that $v/k=td_1$, $t\in\mathbb{Z}$. Since $y=ad_1+i$, we have $y+v/k=ad_1+i+v/k=(a+t)d_1+i$; therefore, $y+v/k$ is a point in U_i. Following the same argument, we can show that if $kg\nmid v$, then $y+v/k$ appears in U_{i+r}, where $v/k=qd_1+r$, $r\neq 0$.

Lemma 5.14 ([9]). *Let $S=(V,\mathscr{B})$ be a cyclic BIBD(v,k,λ), $k\geq 3$ and let $B_d=\{0,d_1,\ldots,d_{(k-1)}\}$ be a non-regular base block of the design. Let $U(B_d,d_1)=\{U_0,U_1,\ \ldots,U_{m-1}\}$, $1\leq m\leq v/2$, be a collection of length-ℓ partial Ucycles representing the blocks of* Dev(B_d)*, with d_1 used to form the partial Ucycles. If B_e is a repeated regular short-orbit base block, then two copies of a block in* Dev(B_e) *can be inserted into the partial Ucycles of $U(B_d,d_1)$.*

Proof. Let S be a cyclic BIBD(v,k,λ) and let $B_d=\{0,d_1,\ldots,d_{(k-1)}\}$ be a full-orbit or non-regular short-orbit base block of the design. Let $U(B_d,d_1)=\{U_0,U_1,\ldots,U_{m-1}\}$ be a collection of length-ℓ partial Ucycles representing the blocks of Dev(B_d), with d_1 used to form the partial Ucycles. We will insert two blocks of the form $x+B_e$, $x\in\{0,1,\ldots,v/k-1\}$. Select points x and $x+v/k$ to represent both copies of $x+B_e$. Note that to represent any block in Dev(B_e), we require only points in $\{0,1,\ldots,2v/k-1\}$. As non-regular short-orbit base blocks generate at least $2v/k$ blocks, each of the points in $\{0,1,\ldots,2v/k-1\}$ appears in the collection of partial Ucycles $U(B_d,d_1)$. There exists $i\in\{0,\ldots,m-1\}$ such that $x\in U_i$ and there exists $j\in\{0,\ldots,m-1\}$ such that $x+v/k\in U_j$. The method used to insert the two blocks depends on the relationship of i to j.

Case 1: If $i=j$, then x and $x+v/k$ appear in the same partial Ucycle of $U(B_d,d_1)$. Use Insertion A (see Fig. 5.3), but do not change the labels of any points (i.e. do not change ad_1 to ad_1+d_s nor cd_1 to cd_1+d_s). We will refer to this type of insertion as **Insertion A'**.

Case 2: If $i\neq j$, then x and $x+v/k$ appear in different partial Ucycles of $U(B_d,d_1)$. Use Insertion B (see Fig. 5.4), but do not change the labels of any points. We will refer to this type of insertion as **Insertion B'**. □

We have four methods for inserting two regular short-orbit blocks into a collection of partial Ucycles at the same time. When we use each of these methods depends on how many copies of the regular short-orbit base block appear in the design in question and the difference used to create the collection of partial Ucycles we have chosen to insert into. We would like to perform the types of insertion described in Lemmas 5.13 and 5.14 multiple times within a given collection of partial Ucycles. As Insertions A' and B' are the most simple insertions, we prefer to insert as many regular short-orbit blocks as possible using these two methods.

Suppose B_e is a regular short-orbit base block in a BIBD(v,k,λ). If $\lambda > 1$ and B_e is repeated, we pair copies of B_e and insert pairs of blocks using Insertions A' and B'. If $\lambda = 1$ or we have one copy of B_e remaining, we pair distinct blocks of Dev(B_e) and insert pairs by Insertions A and B. Notice that if $v/k \equiv 1 \pmod 2$ one block of Dev(B_e) will not be paired. Lemma 5.19 deals with how to insert this single block.

In Figs. 5.3 and 5.4, we illustrated partial Ucycles by writing the points of the Ucycle and connecting each pair of points by an edge. We can think of each edge as representing the block represented by the two endpoints. We will assign a colour to each edge of a partial Ucycle to illustrate the relationship of the two endpoints. Note that these colours do not represent the base block from which the block represented was developed (as in the base block coloured pair adjacency graph); instead, colours will represent the difference between the two endpoints of an edge relative to a fixed direction of reading. Ucycles will always be read clockwise. If, when reading a partial Ucycle in the clockwise direction, the endpoints of the edge increase by d_1 (the difference used to create the Ucycle), colour the edge blue. If the endpoints of the edge decrease by d_1, colour the edge red. When a partial Ucycle is first created all of its edges are blue. However, when we insert a pair of blocks into a collection of partial Ucycles, a segment of a partial Ucycle (or an entire partial Ucycle) is reversed. Suppose we wish to insert a pair of blocks where at least one of the blocks is represented by the pair $\{y,z\}$, $y,z \in \mathbb{Z}_v$. To insert this pair of blocks, a single partial Ucycle (or two partial Ucycles in a collection of partial Ucycles) is broken at the point y and at the point z and two segments are created. One of these segments is reversed and then the two segments are reattached so that two copies of the edge $\{y,z\}$ have been introduced into the partial Ucycle. If the two blocks inserted are distinct, then one of the $\{y,z\}$ edges has endpoint labels changed to $\{y+d_s, z+d_s\}$. We colour the two new edges green and we swap red for blue and blue for red on each edge of the reversed segment. All other edge colours are unchanged.

Lemma 5.15 ([9]). *Let S be a cyclic BIBD(v,k,λ), $k \geq 3$, with a regular short-orbit base block B_e and a non-regular base block $B_d = \{0,d_1,\ldots,d_{(k-1)}\}$. Let $U(B_d,d_1) = \{U_0,U_1,\ldots, U_{m-1}\}$, $1 \leq m \leq v/2$, be a collection of length-ℓ partial Ucycles representing the blocks of* Dev(B_d)*, with d_1 used to form the partial Ucycles. If B_e appears twice in S, then two complete sets of blocks developed from B_e can be inserted into the partial Ucycles of $U(B_d,d_1)$.*

Proof. Let S be a cyclic BIBD(v,k,λ), $k \geq 3$, with full-orbit or non-regular short-orbit base block $B_d = \{0,d_1,\ldots,d_{(k-1)}\}$. Let $U(B_d,d_1) = \{U_0,U_1,\ldots, U_{m-1}\}$ be a collection of length-ℓ partial Ucycles representing the blocks of Dev(B_d), with d_1 used to form the partial Ucycles. Suppose that B_e is a repeated regular short-orbit base block in S. For each $x \in \{0,1,\ldots,v/k-1\} \subset \mathbb{Z}_v$, we will insert the block $x+B_e$ twice into the partial Ucycles of $U(B_d,d_1)$ using insertion method A' or B'. We must ensure that we never wish to insert a regular short-orbit block at a point that has been involved in a previous insertion. Choose $\{0,v/k\}$ to represent B_e. For each $x \in \{0,1,\ldots,v/k-1\}$, the block $x+B_e$ will be represented by $\{x,v/k+x\}$. These pairs are pairwise disjoint; therefore, each block of Dev(B_e) is inserted at a unique pair of points and so two copies of each block in $\mathrm{Dev}(B_e)$ can be inserted into $U(B_d, d_1)$. In visual terms, the disjointness of the insertion pairs implies that we never wish to insert at a point incident to a green edge. □

Corollary 5.4 ([9]). *Let S be a cyclic BIBD(v,k,λ), $k \geq 3$, with regular short-orbit base block B_e. If $n_3 = 2$ and $n_1 + n_2 \geq 1$, then there exists a collection of length-ℓ partial Ucycles representing the blocks developed from a non-regular base block into which all blocks from two copies of* $\mathrm{Dev}(B_e)$ *can be inserted.*

If there are an odd number of regular short-orbit base blocks in a BIBD, we must insert one of the sets of $\mathrm{Dev}(\{0, v/k, \ldots, (k-1)v/k\})$ singly. This will require the use of insertion methods A or B and the existence of a full-orbit base block containing a difference coprime to v/k.

Lemma 5.16 ([9]). *Let S be a cyclic BIBD(v,k,λ), $k \geq 3$. There exist at least*

$$\left\lceil \frac{\lambda\phi(v/k)}{k-1} \right\rceil - n_2$$

full-orbit base blocks each containing a difference coprime to v/k.

Proof. Let S be a cyclic BIBD(v,k,λ), $k \geq 3$. The number of elements in $\mathbb{Z}_v$ that are coprime to v/k is $k \cdot \phi(v/k)$. The total number of such elements appearing as differences in the base blocks of S is $\lambda k\phi(v/k)$. Notice that none of these differences can be in regular short-orbit base blocks; therefore, the minimum number of non-regular base blocks in S containing differences coprime to v/k (differences not necessarily all unique) is

$$\left\lceil \frac{\lambda k\phi(v/k)}{2\binom{k}{2}} \right\rceil,$$

since the number of differences appearing in a base block is $2\binom{k}{2}$. To ensure that x of the base blocks containing a difference coprime to v/k are full-orbit base blocks, we must have $x \leq \lceil(\lambda\phi(v/k))/(k-1)\rceil - n_2$. □

The proof of the Theorem 5.13 will require the existence of full-orbit base blocks containing differences coprime to v (not v/k, as stated above in Lemma 5.16 and in several of the following lemmas). Of course, any difference coprime to v is coprime to v/k; therefore, in the proof of Theorem 5.13, we will appeal to the following corollary.

Corollary 5.5 ([9]). *Let S be a cyclic BIBD(v,k,λ), $k \geq 3$. There exist at least*

$$\left\lceil \frac{\lambda\phi(v)}{k(k-1)} \right\rceil - n_2,$$

full-orbit base blocks each containing a difference coprime to v.

Lemma 5.17 ([9]). *Let S be a cyclic BIBD(v,k,λ), $k \geq 3$, with a regular short-orbit base block B_e. If*

$$x \leq \left\lceil \frac{\lambda\phi(v/k)}{(k-1)} \right\rceil - n_2,$$

and $x \geq 1$, then there exists a collection of length-ℓ partial Ucycles representing the blocks generated by a full-orbit base block into which all blocks of Dev(B_e), *except possibly one, can be inserted without using the partial Ucycles representing a predetermined set of $x-1$ full-orbit base blocks.*

Proof. Let S be a cyclic BIBD(v,k,λ), $k \geq 3$, with $n_2 \leq \lceil(\lambda\phi(v/k))/(k-1)\rceil - x$, for some $x \geq 1$. Lemma 5.16 implies that there exists a full-orbit base block in S, denoted $B_d = \{0, d_1, \ldots, d_{(k-1)}\}$, which is distinct from $x-1$ predetermined full-orbit base blocks and which contains at least one difference coprime to v/k. Let $d_s \in B_d$ be a difference coprime to v/k and let $U(B_d, d_1) = \{U_0, U_1, \ldots, U_{m-1}\}$ be a collection of length-ℓ partial Ucycles representing the blocks of Dev(B_d), with $d_1 \neq d_s$ used to form the partial Ucycles.

Suppose B_e is a regular short-orbit base block in S. The points in the blocks of Dev(B_e) are disjoint; therefore, we will use Insertions A and B to insert blocks of Dev(B_e) into partial Ucycles. We pair the blocks of Dev(B_e) as follows: pair B_e with $d_s + B_e$, next pair $2d_s + B_e$ with $3d_s + B_e$, and so on, where the multiplication $x \cdot d_s$ is taken modulo v/k. As $\gcd(v/k, d_s) = 1$, at most one of the blocks in Dev(B_e) is not paired. We prove that all the pairs can be inserted into $U(B_d, d_1)$.

Prior to insertion, we do not know which pair of points we will use to represent any of the blocks in Dev(B_e); however, because the blocks of Dev(B_e) are disjoint there is no possibility of trying to make an insertion involving a point that has already been used for insertion. Which points are selected to represent the block $2xd_s + B_e$, $x \in \mathbb{Z}$, will be determined by the colours of the edges in $U(B_d, d_1)$ incident to each of the points in $2xd_s + B_e$ at the time of insertion. Since the blocks of Dev(B_e) are disjoint, none of the points in $2xd_s + B_e$ will be incident to a green edge before $2xd_s + B_e$ is inserted. Furthermore, since every insertion creates a green edge between the two segments of partial Ucycles involved in the insertion, every sequence of blue edges is separated from every sequence of red edges by a green edge. That is, blue and red edges are never incident to each other. As a result, each pair of edges incident to each of the points in $2xd_s + B_e$ must either both be blue or both be red. There are at least three unique points in $2xd_s + B_e$, so in $U(B_d, d_1)$ at least two of these points have incident edges of the same colour.

Select two points in $2xd_s + B_e$ that have edges of the same colour incident to them in $U(B_d, d_1)$. Denote the pair $\{ad_1 + i, cd_1 + j\}$, $a, c \in \mathbb{Z}_v$, where $i = j$ if the pair of points is in the same partial Ucycle of $U(B_d, d_1)$ and $i \neq j$ if each point is in a different partial Ucycle of $U(B_d, d_1)$. Without loss of generality, assume $a \leq c$ (a may be equal to c only if $i \neq j$). This pair of points will be used to represent $2xd_s + B_e$. As we are employing insertions A and B, when we insert $2xd_s + B_e$, we will also insert $d_s + 2xd_s + B_e$ and this block will be represented by the pair $\{ad_1 + i + d_s, cd_1 + j + d_s\}$. Recall that the rule for determining which copy of the introduced edge $\{ad_1+i, cd_1+j\}$ will be changed to $\{ad_1+i+d_s, cd_1+j+d_s\}$ is change the ad_1+i adjacent to $(a+1)d_1+i$ to ad_1+d_s+i and change the cd_1+j adjacent to $(c+1)d_1+j$ to cd_1+d_s+j. This change ensures that the block ad_1+i+B_d, which had previously been represented by the edge $\{ad_1+i,(a+1)d_1+i\}$, is still in the partial Ucycle but is now represented by the pair $\{ad_1+i+d_s,(a+1)d_1+i\}$. Similarly, cd_1+j+B_d is now represented by the pair $\{cd_1+j+d_s,(c+1)d_1+j\}$. In colouring terms, this means that we change the $\{ad_1+i, cd_1+i\}$ edge which (when reading the Ucycle in the clockwise direction) has a red edge preceding it and a blue edge following it.

To ensure that multiple insertions can be made into $U(B_d,d_1)$ we must ensure that when we change point labels the edges incident to these changed points continue to represent the same blocks. Our argument is based on the rule used to determine which of the two new green edges has its endpoints changed when performing an insertion of two distinct blocks. The rule says we change the endpoints of a green edge if it has a red edge preceding it and a blue edge following it. Note that this edge colour order (red, green, blue) is independent of reversal of any segment containing these three edges since reversal of a segment swaps the colour of red and blue edges. Furthermore, any segment to be reversed that contains one of these three edges must contain all of these edges.

Without loss of generality, suppose ad_1+i+d_s is the changed point adjacent to the red edge. Then the edge colour-point sequence around the changed edge is red edge, ad_1+i+d_s, green edge, cd_1+j+d_s, blue edge. For the red edge to represent ad_1+i+B_d its other endpoint must be $(a+1)d_1+i$. That is, the other endpoint must not be a changed point. Denote this endpoint by w. If w were a changed point, then the other edge it is adjacent to would be green. However, the change rule says that we change the endpoints of the green edge if it has a blue edge *following* it; therefore, w must not be a changed point. Similarly, since cd_1+j+d_s is the changed point adjacent to the blue edge, the other endpoint of the blue edge must be $(c+1)d_1+j$, in order for the blue edge to continue to represent cd_1+j+B_d. Denote this other endpoint of the blue edge by w'. If w' were a changed point, then the other edge it is adjacent to would be green. However, the change rule says that we change the endpoints of the green edge if it has a red edge *preceding* it; therefore, w' must not be a changed point. We conclude that all paired blocks of B_e can be inserted into $U(B_d,d_1)$ without affecting the blocks represented by the original partial Ucycles in $U(B_d,d_1)$. Figure 5.5 illustrates this argument. The circles represent unknown points that have no impact on the argument. □

We restate Lemma 5.17 in terms of $\phi(v)$, instead of $\phi(v/k)$.

Corollary 5.6 ([9]). *Let S be a cyclic BIBD(v,k,λ), $k \geq 3$, with regular short-orbit base block B_e. If $x \leq \lceil \lambda\phi(v)/(k(k-1)) \rceil - n_2$, and $x \geq 1$, then there exists a collection of length-ℓ partial Ucycles representing the blocks generated by a full-orbit base block into which all blocks of* Dev(B_e)*, except possibly one, can be inserted without using the partial Ucycles representing a predetermined set of $x-1$ full-orbit base blocks.*

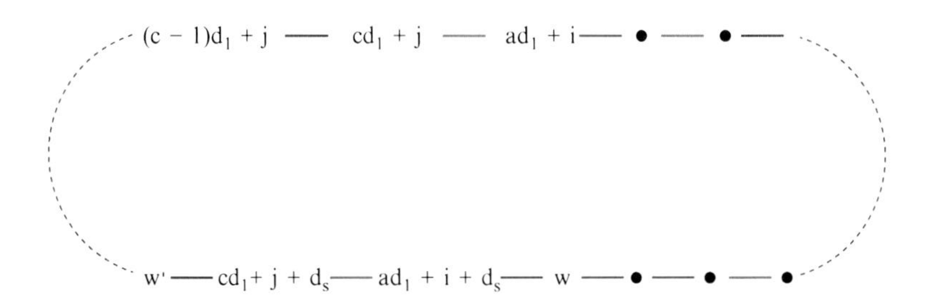

Fig. 5.5 Insertion of multiple pairs of blocks developed from a regular short-orbit base block into a collection of partial Ucycles

With Lemmas 5.15 and 5.17, we have established that all, or all save one, blocks developed from a regular short-orbit base block can be inserted into an existing collection of partial Ucycles. The following two lemmas deal with insertion of a single regular short-orbit block. Note that the type of insertion we describe will need to be executed only if both n_3 and v/k are odd. Lemmas 5.18 and 5.19 rely on these parity assumptions. Before introducing the final type of insertion, Lemma 5.18 establishes when the properties necessary for execution of this insertion hold. To aid our discussion, we will say a difference d is **unusable** if all λ copies of each difference $d \pm r \cdot v/k \pmod{v}$, $r \in \mathbb{Z}_k$, appear in the same non-regular base block.

Lemma 5.18 ([9]). *Let S be a cyclic BIBD(v,k,λ), $k \geq 3$. If the number of regular short-orbit base blocks in S is odd, v/k is odd, and the number of non-regular base blocks in S is at least two, there exist non-regular base blocks $B_d = \{0, d_1, \ldots, d_{(k-1)}\}$ and $B_f = \{0, f_1, \ldots, f_{(k-1)}\}$ such that $f_t - d_s$ is equal to a difference in a regular short-orbit base block for some $s,t \in \{1,2,\ldots,(k-1)\}$.*

Proof. Let S be a cyclic BIBD(v,k,λ), $k \geq 3$, and let B_e be a regular short-orbit base block of the design. We wish to find differences $d_s, f_t \in \mathbb{Z}_v \setminus \{0\}$, with these differences appearing in two different non-regular base blocks, such that $f_t - d_s$ is a difference in B_e. Consider $W = \{0, v/k, 2v/k, \ldots, (k-1)v/k\}$—the differences in a regular short-orbit base block with the addition of $\{0\}$—as a subgroup of $\mathbb{Z}_v$. The cosets W, $W \pm 1 \ldots, W \pm (v-k)/2k$ are disjoint and partition $\mathbb{Z}_v$. If a difference $d_s \in \mathbb{Z}_v \setminus \{0\}$ is unusable, all λ copies of each of the differences in $W - d_s$ and $W + d_s$ must be in the same non-regular base block.

Suppose every difference in $\mathbb{Z}_v$ is unusable. For each difference in a base block, λ copies of the coset containing this difference and λ copies of the coset containing its negative must appear in the base block. The differences in a pair of cosets of the form $W + d$, $W - d$, must completely fit in a single base block and no portion of a coset may appear in another base block. If the number of differences in a non-regular base block is not divisible by $2k\lambda$, or if the cosets in the base block do not account for all differences in the base block, then the remaining differences in the base block are multiples of v/k. We break our analysis into two cases.

Case 1: Suppose $\lambda = 1$. S must contain exactly one regular short-orbit base block and the remaining blocks of the design must be full-orbit base blocks.

The differences in $W \setminus \{0\}$ appear only in the regular short-orbit base block. Furthermore, the cosets $W \pm i$, $i \in \{1,2,\ldots,(v-k)/2k\}$ pack perfectly into the full-orbit base blocks because no multiples of v/k are available to complete the packing. If a full-orbit base block contains an element of $W+d$, $d \in \{1,2,\ldots,(v-k)/2k\}$, the block must contain all differences in $W+d$ and $W-d$. Therefore, since each coset contains k differences and each full-orbit base block contains $2\binom{k}{2}$ differences, we must have $2k|2\binom{k}{2}$, which implies $2|(k-1)$. Thus, k must be odd, and since v/k is also odd, v must be odd. Consider a full-orbit base block B and, without loss of generality, assume $0 \in B$. Let B^d denote the set of differences in B. Since the cosets must pack exactly, $B^d = \{W-x_1, W+x_1, W-x_2, W+x_2, \ldots, W-x_{(k-1)/2}, W+x_{(k-1)/2}\}$, where, for $1 \le i \le (k-1)/2$, $x_i \in \{1,2,\ldots,(v-k)/2k\}$. B cannot have two (or more) elements from the same coset appearing as points, otherwise a difference that is a multiple of v/k would be in B and all such differences must be in the regular short-orbit base block. Taking the points of B^d modulo v/k yields the set $B^d_{\text{mod}} = \{\pm x_1, \pm x_2 \ldots, \pm x_{(k-1)/2}\}$. Each of the points in B^d_{mod} represents a coset, and at most one representative of each coset may appear as a point in B. Since the number of non-zero points in B is $k-1$ and $|B^d_{\text{mod}}| = 2 \cdot (k-1)/2$, a point from each coset in B^d must appear in B.

Let us take a closer look at B^d_{mod}. We show that $B^d_{\text{mod}} \cup \{0\}$ is a subgroup of $\mathbb{Z}_{v/k}$. Zero is in $B^d_{\text{mod}} \cup \{0\}$, and for each $x \in B^d_{\text{mod}}$ we can see that $-x \in B^d_{\text{mod}}$. To see that B^d_{mod} is closed under addition, consider two points, $x_i, x_j \in B^d_{\text{mod}}$. Mapping these elements back to points in B, these elements have the form $x_i + rv/k$, $x_j + r'v/k$, where $r, r' \in \{0,1,\ldots,(k-1)\}$. Now $x_j - x_i \equiv x_j - x_i + (r'-r)v/k \pmod{v/k}$, so an element of the coset $W \pm (x_j - x_i)$ appears as a difference of two points in B. Because all elements of a coset appear in exactly one base block $x_j - x_i \in B^d_{\text{mod}}$. The modulo v/k mapping can be applied to each of the full-orbit base blocks in S; recall that we have assumed there exist at least two such base blocks in S. The result is a non-trivial collection of proper subgroups that together contain each element of $\mathbb{Z}_{v/k} \setminus \{0\}$ exactly once (and each subgroup contains zero). This is impossible since no proper subgroup can contain the generator 1.

Case 2: Suppose $\lambda > 1$. This case breaks up into three additional subcases: first, suppose that all differences that are multiples of v/k are in regular short-orbit base blocks; second, suppose some of the differences $\{v/k, 2v/k, \ldots, (k-1)v/k\}$ are not in regular short-orbit base blocks; and finally, suppose some of the differences $\{v/k, 2v/k, \ldots, (k-1)v/k\}$ are not in regular short-orbit base blocks and that the differences not in regular short-orbit base blocks are all in the same non-regular base block. The details of this case can be found in [9] and are left as an exercise. □

Lemma 5.19 ([9]). *Let S be a cyclic BIBD(v,k,λ), $k \ge 3$, with a regular short-orbit base block B_e. If n_3 is odd, v/k is odd, and $n_1 + n_2 \ge 2$, then there exist non-regular base blocks $B_d = \{0, d_1, \ldots, d_{(k-1)}\}$ and $B_f = \{0, f_1, \ldots, f_{(k-1)}\}$, containing differences d_s and f_t, respectively, such that $f_t - d_s = rv/k$, $r \in \{1,\ldots,(k-1)\}$. Let*

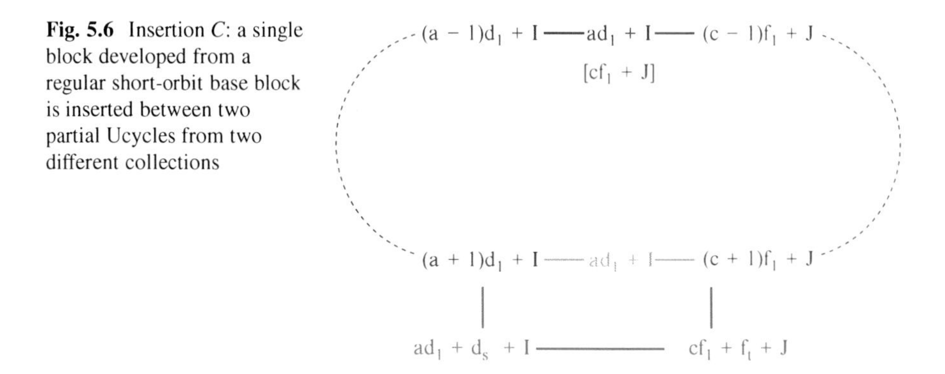

Fig. 5.6 Insertion C: a single block developed from a regular short-orbit base block is inserted between two partial Ucycles from two different collections

$U(B_d,d_1)=\{U_0,U_1,\ldots,U_{m-1}\}$ and $U(B_f,f_1)=\{Q_0,Q_1,\ldots,Q_{m'-1}\}$ be collections of partial Ucycles representing the blocks of $\mathrm{Dev}(B_d)$ *and* $\mathrm{Dev}(B_f)$*, respectively, with $d_1\neq d_s$ and $f_1\neq f_t$ used to form the sequences. The block $x+B_e$, $x\in\{0,1,\ldots\ldots,v/k-1\}$, can be inserted between a partial Ucycle in $U(B_d,d_1)$ and a partial Ucycle in $U(B_f,f_1)$.*

Proof. Let S be a cyclic BIBD(v,k,λ), $k\geq 3$, with regular short-orbit base block B_e. Suppose we wish to insert the block $x+B_e$, for some $x\in\{0,1,\ldots,v/k-1\}$, into existing partial Ucycles representing some of the blocks of S. Since n_3 is odd, and v/k is odd, and the number of non-regular base blocks is at least two, Lemma 5.18 implies there exist two non-regular base blocks, $B_d=\{0,d_1,\ldots,d_{(k-1)}\}$ and $B_f=\{0,f_1,\ldots,f_{(k-1)}\}$, such that $f_t-d_s=rv/k$, $s,t\in\{2,3,\ldots,k-1\}$, for some $r\in\{1,2\ldots,(k-1)\}$. Let $U(B_d,d_1)=\{U_0,U_1,\ldots,U_{m-1}\}$ be a collection of partial Ucycles representing $\mathrm{Dev}(B_d)$, with $d_1\neq d_s$ used to form the sequences. Let $U(B_f,f_1)=\{Q_0,Q_1,\ldots,Q_{m'-1}\}$ be a collection of partial Ucycles representing the blocks of $\mathrm{Dev}(B_f)$, with $f_1\neq f_t$ used to form the sequences.

The block to be inserted, $x+B_e$, will be represented by the pair $\{x,x+rv/k\}$. To perform this insertion we must find $x-d_s$ in one of the partial Ucycles in $U(B_d,d_1)$ and in $U(B_f,f_1)$. If B_d or B_f is full-orbit, this is easy to do; if either is short then the point may not appear, but we can make a constant additive adjustment to each point so that it does appear. We will insert $x+B_e$ between partial Ucycles U_I and Q_J. Break U_I between $(a-1)d_1+I$ and ad_1+I, then add ad_1+I to the end of the segment $ad_1+I,(a+1)d_1+I,\ldots(a-1)d_1+I$ to obtain a segment starting and ending with ad_1+I. Break Q_J between $(c-1)f_1+J$ and cf_1+J, then add cf_1+J to the end of the segment $cf_1+J,(c+1)f_1+J,\ldots,(c-1)f_1+J$ to obtain a segment starting and ending with cf_1+J.

This insertion is illustrated in Fig. 5.6. Green lines and numbers represent changes to the original partial Ucycles—the green lines are new connections and the green numbers are altered numbers (with original and identified numbers in square brackets). The blue numbers and lines represent U_I, and the red numbers and lines represent Q_J. We will call this procedure **Insertion *C***. □

The following corollary is the companion to Corollary 5.4 for $n_3=1$.

Corollary 5.7 ([9]). *Let S be a cyclic BIBD(v,k,λ), $k \geq 3$, with regular short-orbit base block B_e and $n_3 = 1$:*

1. *If v/k is even and $n_2 \leq \lceil \lambda\phi(v/k)/(k-1) \rceil - 1$, then there exists a collection of length-ℓ partial Ucycles representing the blocks developed from a full-orbit base block into which all blocks of* $\mathrm{Dev}(B_e)$ *can be inserted.*
2. *If v/k is odd and $n_2 \leq \lceil \lambda\phi(v/k)/(k-1) \rceil - 3$, then there exists a collection of length-ℓ partial Ucycles representing the blocks developed from a full-orbit base block, and two collections of length-ℓ partial Ucycles representing the blocks developed from two non-regular base blocks (distinct from the former) into which all blocks of* $\mathrm{Dev}(B_e)$ *can be inserted.*

We now have all the insertion procedures necessary to insert regular short-orbit blocks into collections of partial Ucycles. It remains to determine when *all* regular short-orbit blocks in a design can be inserted into collections of partial Ucycles representing the non-regular blocks of the design. To minimize the number of collections of partial Ucycles required for insertion of regular short-orbit blocks, execute as many insertions of the type described in Lemma 5.15 as possible. That is, pair copies of the regular short-orbit base block for insertion. All blocks developed from a pair of regular short-orbit base blocks can be inserted into a collection of partial Ucycles representing the blocks generated by one full or one non-regular short-orbit base block. If there are an odd number of regular short-orbit base blocks in a design, S, before doing any insertions, identify up to three collections of partial Ucycles that facilitate the insertion of blocks generated by a single regular short-orbit base block (as per Corollary 5.7). If there are an odd number of regular short-orbit base blocks in S then a full-orbit base block containing a difference coprime to v/k is required. In addition, if v/k is odd, then two non-regular base blocks (either full orbit or non-regular short orbit) having the property that the difference of an element in each of the base blocks is a multiple of v/k are required. Lemma 5.20 states conditions under which all regular short-orbit blocks can be inserted into existing collections of partial Ucycles.

Lemma 5.20 ([9]). *Let S be a cyclic BIBD(v,k,λ), $k \geq 3$:*

1. *If n_3 is even and $n_1 + n_2 \geq n_3/2$, then all blocks generated by regular short-orbit base blocks can be inserted into the collections of partial Ucycles representing the blocks developed from non-regular base blocks.*
2. *If n_3 is odd, $n_1 + n_2 \geq \lfloor n_3/2 \rfloor + 3$ and $n_2 \leq \lceil \lambda\phi(v/k)/(k-1) \rceil - 3$, then all blocks generated by regular short-orbit base blocks can be inserted into the collections of partial Ucycles representing the blocks developed from non-regular base blocks.*

The following corollary will be used to accommodate additional properties required in the proof of Theorem 5.13. From now on, we will assume we require full-orbit base blocks with a difference coprime to v (instead of v/k) for insertion of regular short-orbit blocks. Furthermore, we will use the term **special** to refer to the insertion methods described in Lemmas 5.13 and 5.19 as these require extra properties of the collections of partial Ucycles into which insertions are made.

Corollary 5.8 ([9]). *Let S be a cyclic BIBD(v,k,λ), $k \geq 3$. If $n_1 + n_2 \geq \lfloor n_3/2 \rfloor + 3$ and $n_2 \leq \lceil \lambda\phi(v)/(k(k-1)) \rceil - 4$, then all blocks generated by regular short-orbit base blocks can be inserted into collections of partial Ucycles representing the blocks developed from non-regular base blocks, and this can done such that a length-v partial Ucycle representing the blocks of a full-orbit base block is not used for special insertions.*

The proof is left as an exercise.

We now have all the tools required to prove Theorem 5.13: Let S be a cyclic BIBD(v,k,λ) with $k \geq 3$. If $n_2 \leq \lceil \lambda\phi(v)/(k(k-1)) \rceil - 4$ and $n_1 + n_2 \geq \lfloor n_3/2 \rfloor + 3$, then S admits a Ucycle of rank two.

Proof (Proof of Theorem 5.13). Let S be a cyclic BIBD(v,k,λ), $k \geq 3$, with the required properties. Corollary 5.8 implies that all blocks generated by regular short-orbit base blocks can be inserted into collections of partial Ucycles representing blocks developed from non-regular base blocks. Furthermore, the special insertions can be executed while avoiding a length-v partial Ucycle representing the blocks of a full-orbit base block.

Specifically, to construct a Ucycle for the blocks of S, first determine the parity of the number of regular short-orbit base blocks in S. If the number of regular short-orbit base blocks is odd and v/k is odd, begin by selecting a pair of non-regular short-orbit base blocks satisfying the requirements of Lemma 5.19. Next, select a full-orbit base block containing a difference coprime to v. Every cyclic BIBD(v,k,λ) with $n_2 \leq \lceil \lambda\phi(v)/(k(k-1)) \rceil - 4$ contains such a block. After selecting the three special base blocks, there remains at least one full-orbit base block containing a difference coprime to v. Denote this block B_p. Construct collections of partial Ucycles for each non-regular base block in S using the methods described in Lemmas 5.8, 5.9, 5.10 and 5.11. Take into account that constructing the collections of partial Ucycles representing the blocks developed from the three special base blocks described above requires extra care in the selection of the difference used for development. Furthermore, develop B_p by a difference coprime to v to create a length-v partial Ucycle; denote this partial Ucycle $U(B_p, p_1)$. Once the collections of partial Ucycles have been created, insert the blocks developed from pairs of regular short-orbit base blocks using the method described in Lemma 5.15, but do not use the collections of partial Ucycles representing the three special base blocks. Note that we *can* use the partial Ucycle $U(B_p, p_1)$. Finally, if a single regular short-orbit base block remains, use the set aside collections of partial Ucycles and apply Lemma 5.17, and possibly Lemma 5.19. At this juncture, all blocks of S are represented in partial Ucycles and each partial Ucycle is of length at least two. It remains to show that these partial Ucycles can be joined together to form a Ucycle for the blocks of S.

The method for joining together two partial Ucycles is as follows. Suppose we have two partial Ucycles $U(B_d, d_1)$ and $U(B_f, f_1)$, both containing a point $x \in \mathbb{Z}_v$. Convert each partial Ucycle to a path having x as one endpoint by breaking one of the edges adjacent to x. Suppose we break $U(B_d, d_1)$ such that the endpoints of the segment are x and y and suppose we break $U(B_f, f_1)$ such that the endpoints of the segment are x and z. Join these two segments together by joining the x in $U(B_d,d_1)$ to the z in $U(B_f,f_1)$ and joining the x in $U(B_f,f_1)$ to the y in $U(B_d,d_1)$. The blocks that were represented by $U(B_d,d_1)$ and $U(B_f,f_1)$ are intact, and there is no possibility of creating a 3-claw with this join procedure.

Each collection of partial Ucycles representing a full-orbit base block contains every element of $\mathbb{Z}_v$ exactly once (refer to Lemmas 5.8 and 5.9). Since $U(B_p,p_1)$ is a length-v partial Ucycle (representing a full-orbit base block), $U(B_p,p_1)$ contains every element of $\mathbb{Z}_v$, and therefore every partial Ucycle shares at least one point with $U(B_p,p_1)$. To create a Ucycle representing all blocks of S, we grow the partial Ucycle $U(B_p,p_1)$ by applying the above join procedure with $U(B_p,p_1)$ and another partial Ucycle not yet joined to $U(B_p,p_1)$, until no such partial Ucycles exist. □

We note that the properties $n_1+n_2 \geq \lfloor n_3/2 \rfloor + 3$ and $n_2 \leq \lceil \lambda\phi(v)/(k(k-1)) \rceil - 4$, required by Theorem 5.13, account for the most degenerate cyclic BIBDs. In most cases these inequalities far exceed the actual number of full-orbit base blocks required for creation of a rank two Ucycle. We now focus on several specific families for which the inequalities of Theorem 5.13 can be greatly refined. All specific values of the Euler totient function used in the following corollaries were obtained from [21]. We begin with triple systems.

Corollary 5.9 ([9]). *Every cyclic TS(v,λ), with $v \geq 7$, admits a Ucycle of rank two.*

Proof. Suppose S is a cyclic TS(v,λ). As non-regular short-orbit base blocks do not exist in triple systems (i.e. $n_2 = 0$), all regular short-orbit blocks in S must be inserted into collections of partial Ucycles representing the blocks developed from full-orbit base blocks. The number of blocks in S is $(v(v-1)\lambda)/6$. Each full-orbit base block generates v of these blocks, while each regular short-orbit base block generates $v/3$ of these blocks; therefore,

$$\frac{v(v-1)\lambda}{6} = vn_1 + \frac{v}{3}n_3, \tag{5.2}$$

which implies that the number of full-orbit base blocks used to generate S is

$$n_1 = \frac{(v-1)\lambda}{6} - \frac{n_3}{3}.$$

As $n_3 \leq \lambda$, $n_1 \geq (v-3)\lambda/6$. We now divide our analysis into three cases.

Case 1: Suppose n_3 is even. Lemma 5.20 implies that if $n_1 \geq n_3/2$, all regular short-orbit base blocks can be inserted into collections of partial Ucycles representing full-orbit blocks. Since $n_1 \geq (v-3)\lambda/6$ and $n_3 \leq \lambda$, $n_1 \geq n_3/2$ holds for all triple systems with $(v-3)\lambda/6 \geq \lambda/2$. Simplifying this expression yields $(v-6)\lambda \geq 0$. Therefore, when $v \geq 6$, all regular short-orbit blocks can be inserted into collections of partial Ucycles. Furthermore, when $v \geq 6$, the element $1 \in \mathbb{Z}_v$ must appear as a difference in a full-orbit base block so there exists a length-v partial Ucycle to which all other partial Ucycles can be joined. We conclude that for this case, every cyclic TS(v,λ), $v \geq 6$, admits a rank two Ucycle.

Case 2: Suppose n_3 is odd and $v/3$ is even. To create a Ucycle for the blocks of S, we require the existence of two full-orbit base blocks each containing a difference coprime to v, or one full-orbit base block with all of its differences coprime to v. In the first case, one block will be used in the application of Lemma 5.17 and the other will be used to create a length-v partial Ucycle to which all other partial Ucycles can be joined. In the second case, note that one of the differences coprime to v can be used to create a length-v partial Ucycle and there will be another difference coprime to v in the block that can be used in the application of Lemma 5.17. The minimum number of full-orbit base blocks containing differences coprime to v is $\lceil \lambda\phi(v)/6 \rceil$, so we require $\lambda\phi(v)/6 \geq 1$ to ensure the existence of a length-v partial Ucycle. This inequality holds for all $v \geq 13$.

Insertion of regular short-orbit blocks proceeds as follows. First insert a single copy of $\mathrm{Dev}(\{0, v/3, 2v/3\})$ into a collection of partial Ucycles representing a full-orbit base block containing a difference coprime to v (using the method described in Lemma 5.17). Second, insert the remaining $n_3 - 1$ copies of $\mathrm{Dev}(\{0, v/3, 2v/3\})$ into $(n_3 - 1)/2$ other full-orbit base blocks. This is done via $(n_3 - 1)/2$ applications of Lemma 5.15. In total, insertions of regular short-orbit blocks require $(n_3 - 1)/2 + 1$ distinct full-orbit base blocks. Since $n_1 \geq (v-3)\lambda/6$ and $n_3 \leq \lambda$, we will have sufficient full-orbit base blocks when $(v-3)\lambda/6 \geq (\lambda - 1)/2 + 1$. This inequality simplifies to $(v-6)\lambda \geq 3$; therefore, sufficient full-orbit base blocks exist for all $v \geq 9$.

We now deal with the values of v ($7 \leq v \leq 12$) not covered by the general argument regarding existence of two full-orbit base blocks each containing a difference coprime to v. Since $v/3$ is even, the only value to consider is twelve. The smallest index for which a cyclic $\mathrm{TS}(12, \lambda)$ exists is two [5]. Each of the integers $1, 5, 7, 11$ is coprime to twelve, and for $\lambda \geq 2$, the λ copies of each of these integers cannot all appear as differences in the same full-orbit base block; therefore, there exist at least two full-orbit base blocks each containing a difference coprime to twelve. We conclude that for this case, every cyclic $\mathrm{TS}(v, \lambda)$, $v \geq 7$, admits a rank two Ucycle.

Case 3: Suppose n_3 is odd and $v/3$ is odd. To create a Ucycle for the blocks of S, we require the existence of four full-orbit base blocks each containing a difference coprime to v, or three full-orbit base blocks where one of these base blocks has all differences coprime to v. In the first case, the four blocks ensure that the selection of two blocks for use in the single block insertion (as described in Lemma 5.19) leaves us with at least two blocks each containing at least one difference coprime to v. One of these blocks will be used in the application of Lemma 5.17 and the other block will be used to create a length-v partial Ucycle to which all other partial Ucycles can be joined. In the second case, the full-orbit base block with all differences coprime to v can be developed by a difference coprime to v (yielding a length-v partial Ucycle) while still fulfilling its other role (either in Lemma 5.17 or 5.19). The minimum number of full-orbit base blocks containing differences coprime to v is $\lceil \lambda\phi(v)/6 \rceil$, so we require $\lambda\phi(v)/6 > 2$ (which actually leaves two full-orbit base blocks all of whose differences must be coprime to v), to ensure existence of these blocks. This inequality holds for all $v \geq 44$.

Insertion of regular short-orbit blocks proceeds as follows. First, insert a single copy of $\mathrm{Dev}(\{0, v/3, 2v/3\})$, less one block, into a collection of partial Ucycles representing a full-orbit base block containing a difference coprime to v (using the method described in Lemma 5.17). Second, insert the block leftover between two collections of partial Ucycles (using the method described in Lemma 5.19). Third, insert the remaining $n_3 - 1$ copies of $\mathrm{Dev}(\{0, v/3, 2v/3\})$ (this is done via $(n_3 - 1)/2$ applications of Lemma 5.15) into $(n_3 - 1)/2$ other full-orbit base blocks. In total, insertions of regular short-orbit blocks require $(n_3 - 1)/2 + 3$ distinct full-orbit base blocks. Since $n_1 \geq (v-3)\lambda/6$ and $n_3 \leq \lambda$, we will have sufficient full-orbit base blocks when $(v-3)\lambda/6 \geq (\lambda - 1)/2 + 3$. This inequality simplifies to $(v-6)\lambda \geq 15$; therefore, sufficient full-orbit base blocks exist for all $v \geq 21$.

We now deal with the values of v ($7 \leq v \leq 43$) not covered by the general arguments above. Since $v/3$ is odd, the only values of v to consider are $9, 15, 21, 27, 33$. For $v \geq 21$ only the condition on the number of full-orbit base blocks containing a difference coprime to v must be checked, that is, we wish to determine when $\lambda\phi(v)/6 > 2$. As there exist Steiner triple systems for $v = 21, 27, 33$, we must determine if $\phi(v) > 12$ for each of these specific values of v. The inequality holds for $v = 27, 33$. When $v = 21$, note that for $\lambda \geq 2$, $\lambda\phi(v)/6 > 2$ holds. Therefore, it remains to prove that every cyclic STS(21) admits a Ucycle of rank two. We know that every cyclic STS(21) contains a regular short-orbit base block (see Lemma 2.1) and three full-orbit base blocks, so the minimal requirement of Theorem 5.13 for the number of full-orbit base blocks is not satisfied. However, despite this fact, our construction method can be used to construct Ucycles for each of the non-isomorphic cyclic STS(21)s. Each of the STS(21)s contains a full-orbit base block with two differences coprime to 21, so we can construct a length-21 partial Ucycle (to which all others can be joined) and also insert pairs of short-orbit blocks into this partial Ucycle. The remaining short-orbit block can be inserted between collections of partial Ucycles representing the blocks developed from the two other full-orbit base blocks. Ucycles for the seven non-isomorphic cyclic STS(21)s (listed in [7]) are given in Table 5.11.

For $v = 9, 15$, we must determine when there exist enough full-orbit base blocks for (1) the insertion of all regular short-orbit blocks and (2) the creation of a length-v partial Ucycle to which all other partial Ucycles can be joined, that is, we wish to determine when both $(v-6)\lambda \geq 15$ and $\lambda\phi(v)/6 > 2$ hold. For $v = 9$, the smallest index for which cyclic triple systems exist is three [5], and since $\phi(9) = 6$, both inequalities hold for all TS$(9, \lambda)$s. For $v = 15$, both inequalities hold for all $\lambda \geq 2$. It remains to show that the two non-isomorphic cyclic STS(15)s admit rank two Ucycles. These are given in Table 5.12.

We conclude that every cyclic TS(v, λ), $v \geq 7$, admits a rank two Ucycle. □

Table 5.11 Ucycles of rank two for the seven non-isomorphic cyclic STS(21)s

Base blocks	Base blocks
$\{0,1,3\},\{0,4,12\},\{0,5,11\},\{0,7,14\}$	$\{0,1,3\},\{0,4,12\},\{0,5,15\},\{0,7,14\}$
Ucycle	**Ucycle**
0,7,6,8,1,20,13,12,11,2,14,5,17,8,20,11,10, 1,13,4,16,7,19,10,9,8,10,3,1,2,18,13,8, 3,19,14,9,4,20,15,10,5,11,4,12,3,15,6, 18,9,0,16,11,6,1,17,12,7,2,3,4,5,19,18, 17,16,15,14,16,2	0,7,6,8,1,20,13,12,11,17,2,8,14,4,11,3, 7,11,15,19,2,6,10,14,18,1,5,9,13,17, 0,4,8,12,16,20,5,11,10,9,3,18,12,6, 0,15,9,8,10,3,1,16,10,4,19,13,7,1,2, 3,4,5,19,18,17,16,15,14,16,2

Base blocks	Base blocks
$\{0,1,5\},\{0,2,10\},\{0,3,9\},\{0,7,14\}$	$\{0,1,5\},\{0,2,10\},\{0,3,15\},\{0,7,14\}$
Ucycle	**Ucycle**
0,7,11,18,14,17,20,2,5,8,11,14,15,16,17,10, 9,8,12,5,1,4,7,10,13,16,19,1,2,3,4,14,3, 13,2,12,1,11,0,18,15,12,9,6,3,9,2,10,20, 9,19,8,18,7,17,6,16,5,15,4,5,6,13,12,11, 15,1,18,19,20	0,7,11,18,14,16,18,20,5,11,17,2,8,14,2, 9,1,3,5,7,9,11,13,15,17,19,0,2,4,6, 8,10,12,14,15,16,17,10,9,3,18,12,6, 0,15,9,8,12,5,1,16,10,4,19,13,7,1,2, 3,4,5,6,13,12,11,15,1,18,19,20

Base blocks	Base blocks
$\{0,1,5\},\{0,2,13\},\{0,3,9\},\{0,7,14\}$	$\{0,1,9\},\{0,2,5\},\{0,4,10\},\{0,7,14\}$
Ucycle	**Ucycle**
0,7,11,18,14,15,16,19,1,4,7,10,13,16,17,10, 9,8,12,5,1,2,5,8,11,14,17,20,2,3,4,17,9, 1,14,6,19,11,3,16,8,0,18,15,12,9,6,3,9, 2,13,5,18,10,2,15,7,20,12,4,5,6,13,12, 11,15,1,18,19,20	0,7,3,20,16,1,15,9,13,6,2,19,20,0,1,2, 18,13,8,3,19,14,9,4,20,15,10,5,0,16, 11,6,1,17,12,7,4,11,3,4,5,6,7,8,9,10, 11,12,13,14,15,16,17,18,19,15,11,17, 10,4,8,12,5,1,18,14,10,16,2,17

Base blocks
$\{0,1,9\},\{0,2,5\},\{0,11,17\},\{0,7,14\}$
Ucycle
0,7,17,2,9,3,19,14,9,4,20,15,10,5,0,16, 11,6,1,17,12,7,4,11,3,4,5,6,7,8,9,10, 11,12,13,14,15,16,17,18,19,20,0,1,2, 18,13,8,3,14,4,15,5,19,8,18,3,17,11, 1,12,2,13,6,16,1,15,9,20,10

Corollary 5.10 ([9]). *Every cyclic STS(v), $v \neq 3$, admits a Ucycle of rank two.*

Proof. There are no cyclic STS(v)s for $v = 4,5,6$. The trivial cyclic STS(3) does not admit a Ucycle of rank two as the minimum number of blocks that can be represented by a rank two Ucycle is two. □

Corollary 5.11 ([9]). *Every cyclic TTS(v) admits a Ucycle of rank two.*

Proof. There do not exist cyclic TTS(v)s for $v = 5,6$. The unique TTS(4) is generated by one base block: $\{0,1,2\}$. A Ucycle of rank two for this design is $U = 0,1,2,3$. The two blocks of the trivial TTS(3) are represented by the Ucycle $U = 0,1$. □

Table 5.12 Ucycles of rank two for the two non-isomorphic cyclic STS(15)s

Base blocks	**Base blocks**
$\{0,1,4\},\{0,7,9\},\{0,5,10\}$	$\{0,1,4\},\{0,2,9\},\{0,5,10\}$
Ucycle	**Ucycle**
0, 7, 6, 9, 4, 1, 2, 3, 4, 5, 0, 14, 2, 12, 9, 10, 11, 12, 13, 8, 11, 1, 14, 6, 13, 5, 12, 4, 11, 3, 10, 2, 9, 1, 8	0, 2, 1, 4, 9, 6, 7, 8, 13, 12, 11, 10, 9, 12, 2, 14, 0, 5, 4, 3, 6, 11, 4, 6, 8, 10, 12, 14, 1, 3, 5, 7, 9, 11, 13

The following corollaries can be obtained using arguments similar to those of Corollary 5.9. The details are left to the reader.

Corollary 5.12 ([9]). *Every cyclic BIBD*$(v,4,\lambda)$, *with* $v \geq 85$, *admits a Ucycle of rank two.*

Corollary 5.13 ([9]). *Every cyclic BIBD*$(v,5,\lambda)$, *with* $v \geq 76$, *admits a Ucycle of rank two.*

Corollary 5.14 ([9]).

1. *Every cyclic BIBD*$(v,k,1)$, *with* $k \geq 3$ *and* $k \nmid v$, *admits a Ucycle of rank two.*
2. *Every cyclic BIBD*$(v,k,1)$, *with* $k \geq 3$, $k|v$, $v \geq (3k-2)k$ *and* $\phi(v) > 2k(k-1)$, *admits a Ucycle of rank two.*

Note that for a given k, there exists $v_0 \in \mathbb{Z}$ such that for all $v \geq v_0$, the inequalities required for Corollary 5.14 (2) must hold.

5.3.2.1 The Construction in Action

The procedure for creating a Ucycle of rank two for the blocks of a cyclic BIBD(v,k,λ) is easily formalized and is shown in Algorithm 5.1. We assume we have a cyclic BIBD(v,k,λ) satisfying the requirements of Theorem 5.13.

To further illustrate the construction, let us build a rank two Ucycle for the cyclic STS(27) generated by the full-orbit base blocks $\{0,1,13\}$, $\{0,2,7\}$, $\{0,3,11\}$, $\{0,4,10\}$ and the regular short-orbit base block $\{0,9,18\}$. In preparation for inserting the block that will be leftover after pairing the blocks in Dev$(\{0,9,18\})$, choose $d_2 = 2$ from $\{0,2,7\}$ and $f_2 = 11$ from $\{0,3,11\}$ so that $f_2 - d_2 = 9 \pmod{27}$. We will develop the base block $\{0,2,7\}$ by 7 and develop the base block $\{0,3,11\}$ by 3 to create collections of partial Ucycles representing the blocks of Dev$(\{0,2,7\})$ and Dev$(\{0,3,11\})$, respectively. Let us insert pairs of blocks developed from the regular short-orbit base block into the partial Ucycle representing Dev$(\{0,4,10\})$. If we develop $\{0,4,10\}$ by 4, then, since $\gcd(10,27/3) = 1$, we can pair and insert all blocks in Dev$(\{0,9,18\})$, save one, into this collection of partial Ucycles. Finally, we choose to develop $\{0,1,13\}$ by 1 since this will create a length-27 partial Ucycle representing Dev$(\{0,1,13\})$.

Algorithm 5.1 The procedure for constructing a rank two Ucycle for the blocks of a cyclic BIBD

- Determine if there will be a single regular short-orbit block remaining to be inserted. This will occur if the number of regular short-orbit base blocks in the design is odd and v/k is odd. Find non-regular base blocks $B_d = \{0, d_1, \ldots, d_{(k-1)}\}$ and $B_f = \{0, f_1, \ldots, f_{(k-1)}\}$ such that $f_t - d_s$ is a multiple of v/k. Create collections of partial Ucycles representing the blocks of $Dev(B_d)$ and $Dev(B_f)$, with $d_1 \neq d_s$ and $f_1 \neq f_t$, respectively, used to form the sequences. Set these collections of partial Ucycles aside.
- If there are an odd number of regular short-orbit base blocks in the design, find a full-orbit base block, distinct from B_d and B_f, that contains a difference coprime to v/k. Denote this base block B_p and denote the difference d_p. Create a collection of partial Ucycles representing the blocks of $Dev(B_p)$, not using d_p for development, and set the collection aside.
- Create a collection of partial Ucycles representing the blocks generated by each of the remaining non-regular base blocks. There is at least one full-orbit base block containing a difference coprime to v, denote this block B_q. Ensure that $Dev(B_q)$ is represented by a length-v partial Ucycle by developing the partial Ucycle representing these blocks by a difference coprime to v.
- Pair the regular short-orbit base blocks and for each pair insert the two sets of blocks developed from the pair into a collection of partial Ucycles.
- If there is a regular short-orbit base block remaining, insert all blocks generated by this base block (or all except one if the number of blocks generated is odd) into the collection of partial Ucycles representing $Dev(B_p)$.
- If there is one remaining regular short-orbit block, insert it between two of the partial Ucycles representing $Dev(B_d)$ and $Dev(B_f)$.
- Join all partial Ucycles to the partial Ucycle representing $Dev(B_q)$ to form a Ucycle for the blocks of the design.

1 — 5 — 9 — 0 — 23 — 19 — 15 — 21 — 12 — 6 — 24 — 20 — 16 — 12 — 8
| |
7 14
| |
16 5
| |
10 26
| |
14 — 18 — 22 — 4 — 10 — 19 — 13 — 17 — 21 — 25 — 2 — 11 — 7 — 3

Fig. 5.7 Insertion of pairs of blocks from Dev($\{0,9,18\}$) into the full-length partial Ucycle representing the blocks of Dev($\{0,4,10\}$)

Figure 5.7 illustrates the insertions of pairs from Dev($\{0,9,18\}$) into the partial Ucycle representing Dev($\{0,4,10\}$). The relationship of endpoints of edges (when read in the clockwise direction) is indicated by colour. A red edge indicates decreasing by 4—the difference used for development—while a blue edge represents increasing by 4. The green edges indicate the inserted blocks, and black edges indicate those edges to which directionality no longer applies because of a label change.

Figure 5.8 illustrates the insertion of the block $\{8,17,26\}$ which is the block of Dev($\{0,9,18\}$) remaining after pair insertion. The red edges indicate a partial

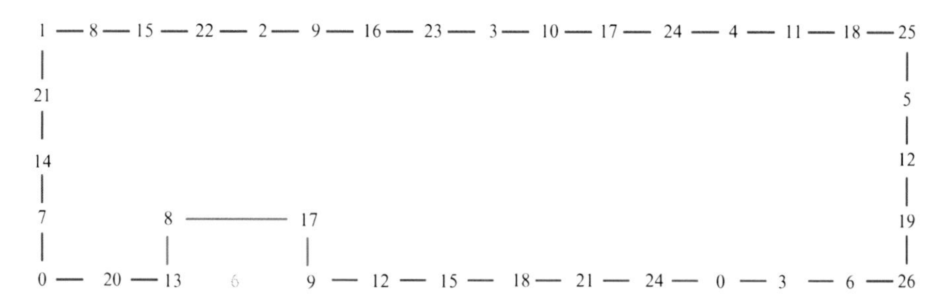

Fig. 5.8 Insertion of the single remaining short orbit block $\{8, 17, 26\}$

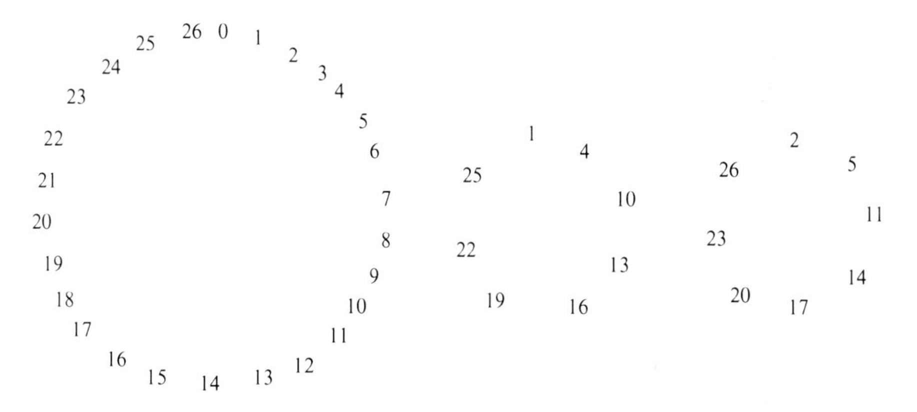

Fig. 5.9 Other (unaltered) partial Ucycles representing the blocks of Dev($\{0, 1, 13\}$) and some of the blocks of Dev($\{0, 3, 11\}$)

Ucycle containing some of the blocks of Dev($\{0,3,11\}$). The blue edges indicate the partial Ucycle representing the blocks of Dev($\{0,2,7\}$). Finally, the green edges are new edges and the gray point indicates the segment altered by the join.

There remain several partial Ucycles untouched by the insertion of regular short-orbit blocks. These partial Ucycles are shown in Fig. 5.9.

All five partial Ucycles (which together contain all blocks of the design) can be joined together to create a Ucycle of rank two for the blocks of this STS(27). There are many ways to join these cycles together. In the following example we have joined all cycles to the partial Ucycle representing Dev($\{0,1,13\}$) (shown in Fig. 5.9 and represented in the Ucycle by bold entries):

$$\mathbf{0}, \mathbf{1}, 4, 10, 13, 16, 19, 22, 25, 1, \mathbf{2}, 5, 11, 14, 17, 20, 23, 26, 2, \mathbf{3}, \mathbf{4}, \mathbf{5}, \mathbf{6}, \mathbf{7},$$

$$\mathbf{8}, 14, 5, 26, 3, 7, 11, 2, 25, 21, 17, 13, 19, 10, 4, 22, 18, 14, 10, 16, 7, 1, 5,$$

$$9, 0, 23, 19, 15, 21, 12, 6, 24, 20, 16, 12, 8, \mathbf{9}, \mathbf{10}, \mathbf{11}, \mathbf{12}, \mathbf{13}, \mathbf{14}, 7, 0, 20,$$

$$13, 6, 26, 19, 12, 5, 25, 18, 11, 4, 26, 8, 0, 3, 6, 9, 12, 15, 18, 21, 24, 17, 10, 3$$

$$23, 16, 9, 2, 22, 15, 8, 1, 21, 14, \mathbf{15}, \mathbf{16}, \mathbf{17}, \mathbf{18}, \mathbf{19}, \mathbf{20}, \mathbf{21}, \mathbf{22}, \mathbf{23}, \mathbf{24}, \mathbf{25}, \mathbf{26}.$$

The construction method presented in this section can be applied to cyclic PBD(v,K,λ)s, with $\min K \geq 3$. All lemmas regarding the creation of collections of partial Ucycles and the insertion of regular short-orbit blocks hold when $\min K \geq 3$. However, what relationship is required between the number of full, non-regular short and regular short-orbit base blocks to ensure all regular short-orbit blocks can be inserted into partial Ucycles and that all partial Ucycles can be joined to form a complete Ucycle for the design is yet to be determined. To ensure there exist sufficient non-regular base blocks for insertion of all regular short-orbit blocks, we must take into account that a cyclic PBD(v,K,λ) may contain several different sizes of regular short-orbit base block. Note that if we have a cyclic PBD(v,K,λ), with $2 \in K$, collections of partial Ucycles for base blocks of size two can be created as long as the base block is not a regular short-orbit base block; a regular short-orbit base block of size two has the form $\{0, v/2\}$ and can only occur when $v \equiv 0 \pmod 2$. A problem may occur because the partial Ucycles created from full-orbit base blocks of size two cannot be used for insertion of a pair of distinct regular short-orbit blocks of any size because there is only one way to represent these size-two blocks, that is, we do not have a d_s in the full-orbit base block that can be added to consecutive points in the partial Ucycle in order to insert two distinct short-orbit blocks (see Lemma 5.13). In addition, we may encounter some difficulty in executing Lemma 5.19 if we discover that the only way to obtain a difference that is a multiple of v/k from points in two different non-regular base blocks involves a base block of size two. To avoid specifying the exact conditions under which the required number of non-regular base blocks occur, we say that a cyclic PBD(v,K,λ) for which all regular short-orbit base blocks can be inserted into collections of partial Ucycles has a **sufficient** number of non-regular base blocks. Then we have the following generalization of Theorem 5.13.

Theorem 5.14 ([9]). *Let S be a cyclic PBD(v,K,λ). If* $\min K \geq 3$ *and if S has a sufficient number of non-regular base blocks (determined as described in the previous paragraph), then S admits a Ucycle of rank two.*

We reiterate that the method given in this section is the only known method for proving the existence of rank two Ucycles for BIBDs, other than the BIBD $(v,k,\binom{v-2}{k-2})$s given by k-subsets of $[v]$ and the symmetric BIBDs discussed in the next section. There may be other, radically different ways to determine existence, and we encourage the interested reader to seek these out.

Finally, recall our conjecture that any local ordering for the blocks of a design can be expressed in terms of configuration ordering. Unfortunately, the expression of rank two Ucycles in the language of configuration orderings is not the most natural way to express these results. In order to specify that we must avoid the 3-claw, we equate rank two Ucycles with configuration orderings where each configuration has three blocks. A rank two Ucycle for a BIBD is equivalent to a $\mathscr{C}$-ordering for the blocks of the BIBD where $\mathscr{C}$ consists of all configurations having three blocks except the 3-claw. For triple systems, $\mathscr{C}$ consists of the configurations shown in Fig. 5.10.

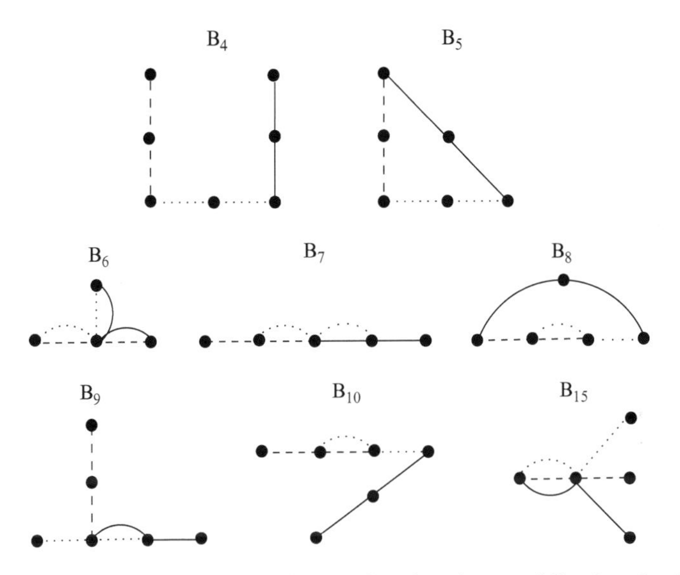

Fig. 5.10 The set of configurations needed to translate the existence of Ucycles of rank two into the language of configuration ordering for triple systems

5.3.3 Universal Cycles of Rank Two for Non-cyclic BIBDs

The methods involved in proving the existence of rank two Ucycles for most cyclic BIBDs are heavily dependent on the structure of the automorphism; however, we would like to be able to determine the existence or non-existence of rank two Ucycles for all BIBD(v,k,λ)s. In this section, we present initial research in this direction.

A family of symmetric designs are the only known type of BIBD for which a construction of Ucycles of rank two that does not utilize a cyclic automorphism group is known. This result complements Corollary 5.3 (see page 143).

The point-block incidence graph, G, of a symmetric BIBD(v,k,λ) is regular of degree k with $|V| = |\mathscr{B}|$. A $\mathscr{B}$-saturating tour (see page 27) in G is equivalent to determining a rank two Ucycle for the design represented. Using this information, Dewar proved that the following family of symmetric designs admit Ucycles of rank two.

Lemma 5.21 ([9]). *Every symmetric BIBD(v,k,λ), with $k \geq 2$ and $v \leq 2k-1$, admits a Ucycle of rank two.*

Proof. The point-block incidence graph of a symmetric BIBD(v,k,λ), whether cyclic or non-cyclic, is a bipartite graph $G = (V, \mathscr{B}, E)$ with $|V| = |\mathscr{B}| = v$, having all vertices of degree k. Create a supergraph of G by adding to G the edges that are required to make $G[\mathscr{B}]$ a complete graph. Denote the resulting graph G'. Note that this new graph is no longer bipartite. To determine when G' admits a Hamilton cycle we appeal to Theorem 2.1 (see page 14). The degree sequence of G' is

$k,\ldots,k,k+v-1,\ldots,k+v-1$; therefore, for $1 \leq i \leq k-1$, we have $d_i > i$. It is clear that $d_i \leq i$, for $k \leq i < 2v/2$; therefore, in order for G' to admit a Hamilton cycle, we require $d_{2v-i} \geq 2v-i$, for all $k \leq i < 2v/2$. That is, we require $k+v-1 \geq 2v-i$, for all $k \leq i < 2v/2$, which simplifies to $v \leq k+i-1$, for all $k \leq i < 2v/2$. Thus, when $v \leq 2k-1$, G' admits a Hamilton cycle. Theorem 2.4 (see page 15) implies that when $v \leq 2k-1$, G is Hamiltonian. Since a Hamilton cycle is certainly a $\mathscr{B}$-saturating tour, such a design admits a Ucycle of rank two. □

Conjectures

As work is just beginning on Gray codes and Ucycles for BIBDs, firm results are minimal and conjectures abound.

As discussed in Sect. 5.2, M. Colbourn and Johnstone's work on the non-existence of 2-intersecting Gray codes for TTS(v)s is incomplete. They describe a construction method for determining twofold triple systems that do not admit minimal change orderings for their blocks; however, they provide only one example of the construction in action. Because the construction involves the assumption that a graph with special properties can be completed to form the 2 block-intersection graph of some TTS(v), it is not clear that the construction can be applied to create a family of graphs having this property.

Conjecture 5.1. There exists $v_0 \in \mathbb{Z}^+$ such that, for all $v \geq v_0$, there exists at least one TTS(v) that does not admit a 2-intersecting Gray code.

The results regarding the existence of Ucycles of rank three for TTS(v)s, $v \equiv 1,4,7 \pmod{12}$, can likely be extended to $v \equiv 10 \pmod{12}$. Recall that Dewar has shown by computer search that no TTS(10) admits a rank three Ucycle.

Conjecture 5.2. For each $v \equiv 10 \pmod{12}$, $v \geq 22$, there exists a TTS(v) that admits a Ucycle of rank three.

We also believe it is possible to "fill in the gaps" in Theorem 5.10.

Conjecture 5.3. For each $v \equiv 1 \pmod{12}$, with $v \equiv 0 \pmod{5}$, there exists a TTS(v) that admits a Ucycle of rank three.

Conjecture 5.4. For each $v \equiv 4 \pmod{12}$, with $v \equiv 0 \pmod{5}$, there exists a TTS(v) that admits a Ucycle of rank three.

Ucycles of rank three cannot exist for TTS(v)s with $v \equiv 0,3,6,9 \pmod{12}$, but for each $v \equiv 3 \pmod{12}$, $v > 3$, there exists a TTS(v) admitting a 2-intersecting cyclic Gray code. Techniques similar to those used in proving this result should be applicable to TTS(v)s with $v \equiv 0,6,9 \pmod{12}$.

Conjecture 5.5. For each $v \equiv 0,6,9 \pmod{12}$, there exists a TTS(v) that admits a 2-intersecting Gray code (or cycle).

Rank two Ucycles do not have as rigid requirements as those of rank three, and we believe such Ucycles exist for broad ranges of parameters and designs. The following conjecture includes proving the details associated to Theorem 5.14 (this particular case is listed as Exercise 5.15).

Conjecture 5.6. Every PBD(v,K,λ), with $\min K \geq 3$ and a sufficient number of non-regular base blocks, admits a Ucycle of rank two.

While we believe this conjecture is true, it is likely that the first steps in the search for Ucycles of rank two for general (not necessarily cyclic) PBDs will begin with proving the following two conjectures.

Conjecture 5.7. Every BIBD$(v,k,1)$, with $k \geq 3$, admits a Ucycle of rank two.

Conjecture 5.8. Every BIBD(v,k,λ), with $k \geq 3$, admits a Ucycle of rank two.

These conjectures and the examples of designs that do not admit Ucycles of rank two indicate that blocks of size two are problematic and must be investigated further.

Exercises and Problems

Exercise 5.1. Complete the details of the proof of Theorem 5.10.

Exercise 5.2. Complete the details of the proof of Theorem 5.11.

Exercise 5.3. Prove Lemma 5.9.

Exercise 5.4. Prove Lemma 5.10.

Exercise 5.5. Prove Lemma 5.11.

Exercise 5.6. Prove Lemma 5.12.

Exercise 5.7. Complete the proof of Lemma 5.13.

Exercise 5.8. Complete the proof of Lemma 5.18.

Exercise 5.9. Prove Corollary 5.7.

Exercise 5.10. Prove Lemma 5.20.

Exercise 5.11. Prove Corollary 5.8.

Exercise 5.12. Prove Corollary 5.12.

Exercise 5.13. Prove Corollary 5.13.

Exercise 5.14. Prove Corollary 5.14.

Exercise 5.15. Complete the details of the proof of Theorem 5.14, that is, for cyclic PBD(v,K,λ)s, with $\min K \geq 3$, determine what relationship is required between the number of full, non-regular short and regular short-orbit base blocks to ensure all regular short-orbit blocks can be inserted into partial Ucycles and that all partial Ucycles can be joined to form a complete Ucycle for the design.

Problem 5.1. Construct tight SCCD(v,k)s with inner or outer expansion sets for $k > 4$.

Problem 5.2. Construct a SCND$(k+1,k)$ or show that such a design cannot exist.

Problem 5.3. Given their similarity to Ucycles, it is natural to consider **cyclic** sequential covering designs. A CSD(v,k,t) can be defined as a cyclic sequence of length t of elements from a v-set such that every pair of elements occur somewhere in the sequence no more than $k-1$ positions apart. Derive some lower bounds and constructions for cyclic sequential covering designs.

Problem 5.4. Find new examples of cubic 3-connected graphs that can be realized as the 2 block-intersection graph of a TTS.

Problem 5.5. Characterize which cubic 3-connected graphs can be realized as the 2 block-intersection graph of a TTS.

Problem 5.6. Show that there exists a function, $f(v) = av + b$ such that for any simple partial TTS(v), and for all $x \geq f(v)$, there exists a simple TTS(x) that contains the given TTS(v) as a subsystem.

Problem 5.7. Show that for all $v \equiv 0,1 \pmod 3$ there exists a TTS(v) whose 2 block-intersection graph is not Hamiltonian. One natural way to approach this problem is to first prove the statement of Problem 5.6 and then apply this result to the partial TTS of M. Colbourn and Johnstone presented in Sect. 5.2.

Problem 5.8. Determine whether, given a graph G which is cubic, 3-connected and does not contain a Hamilton cycle, there exists a graph H such that the construction given on page 123 creates the 2 block-intersection graph of a TTS.

Problem 5.9. In constructing rank three Ucycles and 2-intersecting cyclic Gray codes for TTS(v)s, the concept of a difference sequence was defined. Do difference sequences (cyclic or non-cyclic) exist for all cyclic TTS(v)s?

Problem 5.10. Determine if 2-intersecting cyclic Gray codes exist for TTS(v)s where $v \equiv 0,6,9 \pmod{12}$.

Problem 5.11. Determine if rank three Ucycles exist for TTS(v)s where $v \equiv 10 \pmod{12}$.

Problem 5.12. In Sect. 2.3.1, we noted that for standard binary reflected Gray codes there exist formulas for ranking and unranking, that is, it is possible to determine where a given string appears in the Gray code and, given a position, it is possible to determine which string appears in that position in the list. Ranking and unranking is also possible for many combinatorial Gray codes; however, it is not immediately obvious how to perform ranking and unranking for a κ-intersecting Gray code for a design. This is due, in part, to the fact that the method given for constructing 2-intersecting cyclic Gray codes for cyclic TTS(v)s is not recursive. Do there exist ranking and unranking functions for 2-intersecting cyclic Gray codes for cyclic triple systems?

Problem 5.13. Theorem 5.12 identifies an infinite family of PBDs that do not admit Ucycles of rank two. In addition, there exist the BIBD$(2k,2,1)$s and a PBD$(6,\{2,3\},1)$ that do not admit Ucycles of rank two. Are there other families of designs that do not admit rank two Ucycles?

Problem 5.14. Consider the application of the construction method for Ucycles of rank two, given in Sect. 5.3.2, to constructing $\{B_4,B_5\}$-cyclic orderings.

Problem 5.15. Robert Jamieson has asked if rank two Ucycles for cyclic BIBD (v,k,λ)s can be balanced. More specifically, in a rank two Ucycle for a BIBD (v,k,λ), $S=(V,\mathscr{B})$, does each $v \in V$ appear an equal (or as close as possible) number of times? This is analogous to the question of existence of balanced binary Gray codes; however, in the case of Gray codes, we look at the transition sequence of the Gray code, not at the contents of the Gray code itself (which is certainly balanced). We believe the answer to this question is likely yes. The construction method for partial Ucycles of rank two ensures balance initially. The insertion of regular short-orbit blocks may disrupt this balance; however, we hypothesize that it is possible to maintain balance through careful selection of insertion points.

Problem 5.16. Are there other natural ways to define Ucycles for block designs? In particular, are there other choices for rank besides block size and strength that yield interesting and general results?

References

1. Aldred, R.E.L., Bailey, R.A., McKay, B.D., Wanless, I.M.: Circular designs balanced for neighbours at distances one and two. Draft Manuscript (2011)
2. Batzoglou, S., Istrail, S.: Physical mapping with repeated probes: the hypergraph superstring problem. J. Discrete Algorithms (Oxf.) **1**(1), 51–76 (2000)
3. Buck, M., Wiedemann, D.: Gray codes with restricted density. Discrete Math. **48**, 163–171 (1984)
4. Colbourn, C.J., Colbourn, M.J., Harms, J.J., Rosa, A.: A complete census of (10,3,2) block designs and of Mendelsohn triple systems of order 10. III. (10,3,2) block designs without repeated blocks. Congr. Numer. **37**, 211–234 (1983)
5. Colbourn, C.J., Rosa, A.: Triple Systems. Oxford Mathematical Monographs. The Clarendon Press Oxford University Press, New York (1999)
6. Colbourn, M.J., Johnstone, J.K.: Twofold triple systems with a minimal change property. Ars Combin. **18**, 151–160 (1984)
7. Colbourn, M.J., Mathon, R.A.: On cyclic Steiner 2-designs. Ann. Disc. Math. **7**, 215–253 (1980)
8. Constable, R.L., Phillips, N.C.K., Porter, T.D., Preece, D.A., Wallis, W.: Single-change neighbor designs. Austral. J. Combin. **11**, 247–258 (1995)
9. Dewar, M.: Gray codes, universal cycles and configuration orderings for block designs. Ph.D. thesis, Carleton University, Ottawa, ON (2007)

10. Eades, P., Hickey, M., Read, R.C.: Some Hamilton paths and a minimal change algorithm. J. Assoc. Comput. Mach. **31**(1), 19–29 (1984)
11. Eades, P., McKay, B.: An algorithm for generating subsets of fixed size with a strong minimal change property. Inform. Process. Lett. **19**(3), 131–133 (1984)
12. Goddyn, L., Gvozdjak, P.: Binary Gray codes with long bit runs. Electron. J. Combin. **10**, Research paper 27, (electronic, 10 pp.) (2003)
13. Hare, D.R.: Cycles in the block-intersection graph of pairwise balanced designs. Discrete Math. **137**, 211–221 (1995)
14. Jackson, B.W.: Universal cycles of k-subsets and k-permutations. Discrete Math. **117**, 141–150 (1993)
15. McSorley, J.P.: Single-change circular covering designs. Discrete Math. **197/198**, 561–588 (1999)
16. Phillips, N.C.K.: Finding tight single-change covering designs with $v = 20, k = 5$. Discrete Math. **231**, 403–409 (2001)
17. Phillips, N.C.K., Preece, D.A.: Tight single-change covering designs with $v = 12, k = 4$. Discrete Math. **197/198**, 657–670 (1999)
18. Phillips, N.C.K., Wallis, W.D.: Persistent pairs in single change covering designs. In: Proceedings of the Twenty-fourth Southeastern International Conference on Combinatorics, Graph Theory, and Computing (Florida Atlantic University, Boca Raton, FL, Feb 22–26, 1993), Utilitas Mathematica Publishing, Inc., Winnipeg, MB, Congr. Numer. **96**, 75–82 (1993)
19. Preece, D.A.: Tight single-change covering designs—the inside story. Bull. Inst. Comb. Appl. **13**, 51–55 (1995)
20. Preece, D.A., Constable, R.L., Zhang, G., Yucas, J.L., Wallis, W.D., McSorley, J.P., Phillips, N.C.K.: Tight single-change covering designs. Utilitas Math. **47**, 55–84 (1995)
21. Sloane, N.J.A.: Table of n, phi(n) for $n = 1 \ldots 10000$. http://www.research.att.com/~njas/sequences/b000010.txt (2007). Accessed May 2007
22. Van Rees, G.H.J.: Single-change covering designs II. Congr. Numer. **92**, 29–32 (1993)
23. Vickers, V.E., Silverman, J.: A technique for generating specialized Gray codes. IEEE Trans. Comput. **29**(4), 329–331 (1980)
24. Wallis, W., Yucas, J.L., Zhang, G.H.: Single-change covering designs. Des. Codes Cryptogr. **3**, 9–19 (1992)
25. Wallis, W.D.: Sequential covering designs. In: Graph Theory, Combinatorics, and Algorithms, (Kalamazoo, MI, 1992), Wiley-Interscience Publications, vol. 1–2, pp. 1203–1210. Wiley, New York (1995)
26. Wilf, H.S.: Combinatorial algorithms: an update. CBMS-NSF Regional Conference Series in Applied Mathematics, 55. Society for Industrial and Applied Mathematics (SIAM), Philadelphia, PA (1989)

Chapter 6
Applications

In this chapter, we look at applications of orderings blocks of designs. We will look at erasure-correcting codes, tournament scheduling, reliability testing, group testing and design of specific statistical experiments. Note that terminology and notation of the originating authors is retained, resulting in some inconsistency.

6.1 Erasure-Correcting Codes

Configuration orderings were initially introduced by Cohen and Colbourn in relation to erasure-correcting codes [9]. Parallel disk architectures are a common feature in today's computing environment. These systems, known as redundant arrays of independent disks (RAIDs), connect many disks together in an attempt to improve I/O performance. While connecting several disks increases I/O performance, it has a detrimental effect on the reliability of the system—more disks in the system increases the likelihood that some part of the system (a disk) will fail. When a disk suffers a catastrophic failure, its data is deemed unreadable; therefore, such a disk failure is called an **erasure**. Erasure-correcting codes are designed to recover the erased data when the location of the erasure is known.

The two key factors considered in the evaluation of RAID systems are performance and reliability. The performance of a system is usually judged by examining system response time. The reliability of a system refers to the ability of the system to restore the contents of erased disks. The reliability of a parallel system of disks can be improved by introducing redundancy. As this redundancy often takes the form of parity calculations, the performance of the system is adversely affected; thus, when designing these redundant operations, it is important to reduce the negative impact on performance. An **$[n,c,k]$-erasure-correcting code** consists of an encoding algorithm ε and a decoding algorithm δ with the following properties. Given an n-tuple of data, S (for n disks), ε produces an $(n+c)$-tuple $\varepsilon(S)=(\varepsilon_1(S),\varepsilon_2(S),\dots,\varepsilon_{n+c}(S))$ called a **codeword**, such that for any set $U\subseteq\{1,2,\dots,n\}$, where $|U|=n+c-k$,

M. Dewar and B. Stevens, *Ordering Block Designs: Gray Codes, Universal Cycles and Configuration Orderings*, CMS Books in Mathematics,
DOI 10.1007/978-1-4614-4325-4_6, © Springer Science+Business Media New York 2012

the decoding algorithm δ is able to recover S from $(U, \{\varepsilon_i(S) \colon i \in U\})$ [8]. We will refer to ***k*-erasure-correcting codes** when we do not wish to specify n and c. An $[n,c,k]$-erasure-correcting code that produces codewords in which the first n bits are simply S (the unmodified original data) is called **systematic**. The remaining bits are called the **checks**. We consider only systematic codes in this section. To apply an $[n,c,k]$-erasure-correcting code to an array of disks, the array of disks is broken into two groups by designating the first n disks to be information disks and others to be check disks. **Information disks** contain the original, unmodified data. **Check disks** contain an encoding of the original data determined by parity operations. The above definition for an $[n,c,k]$-erasure-correcting code indicates that the system can tolerate k erasures. To determine an upper bound on k, suppose that $c+1$ disks fail. If these disks are the c check disks and one of the n information disks, it is impossible to recover the erased information. Therefore, we must have $k \leq c$.

In order to discuss the performance of a disk array we introduce the following terms. The **check disk overhead** is the ratio of the number of check disks to the number of information disks. The **update penalty** is the number of check disks whose contents must be changed when a minimal change is made to the contents of a given information disk. For an erasure-correcting code to tolerate k erasures, every update must affect the contents of at least $k+1$ disks (one information disk and k check disks); therefore, the update penalty of a k-erasure-correcting code is at least k. Finally, the **group size** is the number of disks that must be accessed during reconstruction of the contents of a single failed disk. We wish each of these values to be as small as possible. Because update penalties are most important to the performance of a code, we will assume from now on that we are looking at k-erasure-correcting codes with update penalty k.

An $[n,c,k]$-erasure-correcting code can be represented as a $c \times (n+c)$ matrix over $\mathbb{F}_2$. Suppose $H = [P|I]$ is the matrix for an $[n,c,k]$-erasure-correcting code. P is the $c \times n$ matrix that determines the relationship between the check and information disks, and I is the $c \times c$ identity matrix. We identify the columns of P with the n information disks and the columns of I with the c check disks. This matrix is called the **parity check matrix** of the code.

Let us examine a particular erasure-correcting code in order to solidify the ideas introduced thus far. The **full-2 code**, represented by the $c \times (\binom{c}{2} + c)$ matrix $H_{\text{full2}} = [P_{\text{full2}}|I]$, is a 2-erasure-correcting code. The columns of P_{full2} are all possible distinct c-tuples of weight two. For any positive integer k, we can define a full-k code in a similar fashion. H_{full2} for $c = 5$ is given below

$$\boldsymbol{H_{\text{full2}}} = \begin{bmatrix} 1&1&1&1&0&0&0&0&0&0&1&0&0&0&0 \\ 1&0&0&0&1&1&1&0&0&0&0&1&0&0&0 \\ 0&1&0&0&1&0&0&1&1&0&0&0&1&0&0 \\ 0&0&1&0&0&1&0&1&0&1&0&0&0&1&0 \\ 0&0&0&1&0&0&1&0&1&1&0&0&0&0&1 \end{bmatrix}.$$

The first ten columns are associated with the ten information disks, while the last five columns are associated with the five check disks. The contents of check disk i are an encoding of the contents of the information disks whose corresponding columns in P_{full2} have a one in row i. For example, the one in the 11th column of H_{full2} means that the first check disk stores an encoding of the data in the first, second, third and fourth information disks. Note that each column has weight two meaning that each information disk has its data encoded on two check disks. The full-2 code is 2-erasure correcting because if any two disks fail, the data may be restored. However, if three disks fail (e.g. the first information disk and the first two check disks), the data cannot be recovered. In general, the full-2 code has a check disk overhead of $c/\binom{c}{2} = 2/(c-1)$, a group size of $c-1$, and an update penalty of 2.

As noted earlier, update penalties are the primary performance cost associated to an erasure-correcting code. When the update penalty is assumed to be minimized, the next greatest factor in diminishing performance is due to writing data to consecutive information disks. This involves updating some number of check disks. Let u_s denote the average number of check disks updated when s consecutive information disks are written to. Cohen and Colbourn have focused on minimizing u_s through ordering information disks within established systems.

The full-2 code with c check disks can be represented by the complete graph K_c. Each vertex of K_c represents a check disk, and the edges represent information disks (you can read these edges off the columns of H_{full2}). Hellerstein et al. have shown that a code with minimum update penalty two is obtained by having each information disk "checked" by two check disks, with no two information disks sharing the same pair of check disks [27]. Therefore, any 2-erasure-correcting code can be represented by a graph with vertices representing check disks and an edge connecting two check disks representing the information disk they "check". Let $G = (V,E)$ be a graph with m edges. Let d be a fixed positive integer less than m which represents the number of disks involved in a typical write; call d the **window**. An **edge ordering** of G is a permutation of the edge indices $\{0,1,\ldots,m-1\}$. Suppose G is a graph with edge ordering π and window d. For $0 \le i \le m-d$, define graphs $G_i^{\pi,d}$ to be the subgraphs of G induced by the edges $\{e_{\pi(i)}, e_{\pi(i+1)}, \ldots, e_{\pi(i+d-1)}\}$. The **cost** of updating the disks represented by the subgraph induced by d consecutive edges is equal to the sum of the number of edges and the number of vertices of non-zero degree in the subgraph. As d (the number of edges) is fixed, it is clear that the only way to reduce this access cost is to reduce the number of vertices induced. Let $n_i^{\pi,d}$ represent the number of vertices of non-zero degree in $G_i^{\pi,d}$ and define the ***d*-access cost** of graph G under the ordering π to be

$$\frac{\sum_{i=0}^{m-d} n_i^{\pi,d}}{(m+1-d)}.$$

When an ordering π minimizes the d-access cost over all edge orderings of G, it is called a ***d*-optimal ordering** for G. When an ordering π maximizes the d-access cost over all edge orderings of G, it is called a ***d*-pessimal ordering** for G.

The task of reducing the number of check disks accessed in a typical write becomes a problem of reducing the number of vertices induced by any d consecutive edges of an edge-ordering. For example, if $d = 3$, the minimum number of vertices induced by three edges is three, which occurs when the edges form a triangle. The maximum number of vertices induced by three edges is six, which occurs when all edges are disjoint. Therefore, to obtain a 3-optimal ordering, we would like to find an edge ordering where every consecutive set of three edges induces a triangle. This is not possible, unless there are only three vertices in the graph, because the graph does not have repeated edges. As triangles cannot be consecutively induced, the best we can hope for is to alternate between inducing three vertices and inducing four vertices. Whether or not this is possible is determined by the answer to the following question: when can the edges of the complete graph K_c be ordered by a permutation π so that amongst the $\binom{c}{2} - 2$ subgraphs $G_i^{\pi,3}$, at least

$$\frac{1}{4}(c^2 - c - 6) = \frac{1}{2}\left[\left(\binom{c}{2} - 2\right) - 1\right]$$

subgraphs form triangles? Such an ordering of edges is called a **ladder ordering of pairs**. A ladder ordering is called **circular** if for any three consecutive edges, e_i, e_{i+1}, e_{i+2}, the subgraph induced by these edges alternates between having three vertices and having four vertices, with addition of subscripts performed modulo $\binom{c}{2}$. A circular ordering can only exist when $c \equiv 0, 1 \pmod 4$, which ensures that two consecutive triangles are not induced.

Theorem 6.1 ([10]). *A ladder ordering of pairs for K_c exists for all admissible c, except possibly when $c \in \{15, 18, 22\}$.*

A variation on edge ordering is introduced by Cohen et al. in [11]. The ***d*-maximum access cost** of a graph G under edge ordering π is defined to be

$$\max_{0 \leq i \leq m-d} (n_i^{\pi,d}).$$

An edge ordering is **(d, f)-cluttered** if it has d-maximum access cost equal to f. Again, ordering the columns of the parity check matrix in accordance with a cluttered ordering of the corresponding graph minimizes the maximum number of check disks affected by a write operation. Cluttered ordering results were presented in Sect. 4.1. For further details on the existence of cluttered orderings for the complete graph, see [11].

Müller et al. studied cluttered orderings for the complete bipartite graph in [35]. Their motivation to study the complete bipartite graph is the **2-dimensional parity code**, an erasure-resilient code proposed by Hellerstein et al. [27]. This code is defined for $n = \ell^2$ information disks. These disks are imagined to be arranged in an $\ell \times \ell$ array. A check disk is associated with each row and each column and is assigned the parity of that row or column. Note that, contrary to its name, this code is an ℓ^2-dimensional vector space over $\mathbb{Z}_2$, where $\ell^2 > 2$. The name is actually derived from the geometric assignment of check bits to information bits. The 2-dimensional parity code can be represented by the complete bipartite graph, $K_{\ell,\ell}$. The code is 2-erasure-correcting and can correct all three erasures except those that involve a single information disk and its two associated check disks. The code also has optimal check disk overhead.

Similar to the way a graph induces a 2-erasure-correcting code, a Steiner triple system (STS) induces a 3-erasure-correcting code. The triples represent information disks, and the points represent check disks. Thus, an information disk is represented by the check disks associated to it. Such an arrangement corrects all three erasures. It also corrects all four erasures except those corresponding to a single information disk and the three check disks assigned to "check" this information disk.

Cohen and Colbourn have investigated the impact of various (configuration) orderings of the blocks of STSs on the performance of RAID systems. Their results were reported in Sect. 4.1. Theorems 4.2 and 4.3 (see p. 78) provide some insight into bad orderings for the columns of the parity check matrix; however, Cohen and Colbourn are most interested in improving the performance of disk arrays. A good ordering minimizes the number of check disks associated with ℓ consecutive triples, that is, minimizes the number of points induced by ℓ consecutive triples. When $\ell = 2$, an ordering in which pairs of consecutive triples induce the $(5,2)$-configuration consisting of two intersecting triples (see configuration A_2 in Fig. 1.1 on p. 4) is called an **optimal ordering**. Thus, the fact that the 1 block-intersection graph of an $\mathrm{STS}(v)$ is Hamiltonian (see Theorem 4.4 on p. 79) implies that every $\mathrm{STS}(v)$ admits a cyclic optimal ordering.

When $\ell = 3$, a B_5-ordering is an optimal ordering (see Fig. 1.2 on p. 5). The Fano plane (shown in Fig. 2.2 on p. 17) admits a B_5-cyclic ordering: $\{0,1,3\},\{1,2,4\}$, $\{0,2,6\},\{1,5,6\}$, $\{2,3,5\}$, $\{3,4,6\},\{0,4,5\}$. On the other hand, Cohen and Colbourn have shown that there exist $\mathrm{STS}(v)$s of all orders greater than 15 which are not B_5-orderable (see Theorem 4.28 on p. 101). However, they also conjecture that for all $v \geq 15$, there exist B_5-orderable $\mathrm{STS}(v)$s. In [9] they go some way to proving this conjecture by establishing the existence of some small B_5-orderable STSs.

6.2 Tournament Scheduling

A basic form of tournament scheduling requires that, given a set of competitors, every pair from the set must play one game against each other, and a time must be assigned to each game. This can be done by arranging all the pairs from a set of competitors into an array where columns represent the various time slots. To ensure no competitor plays two games at the same time, we require that no competitor appears more than once in each column. This model can be extended to add additional constraints involving venues, rest intervals, home and away games and other relevant factors.

Table 6.1 Three tournament designs: TD(8,4), TD(8,2) and TD(7,3)

0,1	0,2	0,3	0,4	0,5	0,6	0,7
2,6	1,7	1,2	1,3	1,4	1,5	1,6
3,5	3,6	4,6	2,7	2,3	2,4	2,5
4,7	4,5	5,7	5,6	6,7	3,7	3,4

0,1	0,2	0,3	0,4	0,5	0,6	1,6
2,6	3,6	1,2	1,3	1,4	1,5	2,5
3,5	4,5	4,6	5,6	2,3	2,4	3,4

0,1	0,2	0,3	0,4	0,5	0,6	0,7	3,5	3,6	4,6	2,7	2,3	2,4	2,5
2,6	1,7	1,2	1,3	1,4	1,5	1,6	4,7	4,5	5,7	5,6	6,7	3,7	3,4

Definition 6.1 (tournament design). A tournament design, TD(v,c), is a $c \times v(v-1)/2c$ array with entries consisting of the unordered pairs from a v-set, V, having the property that all unordered pairs appear exactly once and every element of V appears at most once in each column.

Typically, rows represent courts where matches are played and columns represent time slots, otherwise known as **rounds**, when matches are played. When a competitor does not play in a round (i.e. does not appear in a column), that competitor is said to have a **bye** round. A tournament design, TD(v,c), is equivalent to a decomposition of K_v into partial 1-factors which contain exactly c disjoint edges (i.e. into c-matchings). A block of a tournament design is the set of matches played in a round. In other words, blocks are the sets of partial 1-factors. Folkman and Fulkerson showed that the necessary conditions for the existence of a tournament design,

$$c \Big| \binom{v}{2}, \tag{6.1}$$

$$1 \leq c \leq \lfloor v/2 \rfloor, \tag{6.2}$$

$$v - 1 + (v \pmod 2) \leq \binom{v}{2} / c, \tag{6.3}$$

are sufficient [23]. Three examples of tournament designs are given in Table 6.1.

Tournament designs can be generalized to matches of size $k > 2$, where k competitors compete together; the resulting designs are called **generalized tournament designs**, GTD(v,c,k). Examples of competitions where matches are of size greater than two are bowling, golfing and racing.

In tournament designs it is often desirable to balance various parameters in order to ensure notions of fairness. These could include distributing the play across better and worse venues, giving all competitors equitably distributed rest intervals and balancing home and away games to mediate "home court advantage". Notions of fairness can require very complex orderings; for example, one might try to evenly distribute the statistical effect of a competitor's previous opponent on their next one. These specialized tournament designs are discussed in the following subsections. Note that there are other forms of balance for tournament designs that do not involve ordering the blocks (rounds). We do not discuss these here, but refer the interested reader to the *CRC Handbook of Combinatorial Designs* [19].

Table 6.2 A court-balanced tournament design, CBTD$(10,3)$

0,1	1,5	2,6	2,7	3,8	3,9	2,3	0,5	4,6	5,7	6,8	7,9	0,8	1,4	4,9
3,6	3,7	1,8	1,9	2,4	0,2	5,8	6,9	1,3	4,8	5,9	0,6	4,7	2,5	0,7
2,8	2,9	3,4	3,5	1,6	1,7	6,7	7,8	8,9	0,9	0,4	4,5	1,2	0,3	5,6

6.2.1 Court-Balance

It is relatively easy to address the issue of venue quality. This is the only specialized tournament design we introduce that does not have a notion of ordering associated.

Definition 6.2 (court-balanced tournament design). A court-balanced tournament design, CBTD(v,c), is a TD(v,c) such that every element of V appears the same number of times, $\alpha = (v-1)/c$, in each row.

The necessary conditions are the same as those for tournament designs with the additional requirement that α be a positive integer. Rodney showed that these necessary conditions are sufficient [41]. A CBTD$(10,3)$ is given in Table 6.2.

In a graphical representation, courts can be modelled by colouring the edges of the underlying complete graph (see Theorem 4.38 on p. 108).

6.2.2 Rest Intervals Between Games

None of the definitions of tournament design introduced thus far involves any explicit ordering; however, there are several different motivations for ordering columns. The first is to balance the interval of rest that each competitor gets between matches. Let T be a TD(v,c) and let $x \in V$. Define $R_T(x)$ to be the minimum number of consecutive columns in which x does not appear, and let S_T be the minimum of these over all elements of V; call this the **separation** of the design.

Lemma 6.1 ([41]). *Let T be any TD(v,c). Define*

$$s(v,c) = \begin{cases} \max\left(0, \left(\frac{v}{2c} - 2\right)\right) & \text{if } v = 0 \pmod{2c} \\ \left\lfloor \frac{v}{2c} \right\rfloor - 1 & \text{otherwise.} \end{cases}$$

Then $S_T \le s(v,c)$.

Proof. The necessary conditions for existence of a TD(v,c) require that $v \ge 2c$. If $2c \le v \le 4c$, then more than half of the elements appear in each column and so the maximum separation must be zero (i.e. $S_T = 0$), and it is easy to see that $S_T \le s(v,c)$.

Let $v > 4c$ and suppose S_T is attained by $w \in V$. Consider the closest two appearances of w in columns c_1 and $c_2 > c_1$ with $S_T = c_2 - c_1 - 1$. By definition each column has no repeated elements, and since no element has separation less than S_T, we know that all columns, i, $c_1 \le i < c_2$, must be disjoint. The subarray of

Table 6.3 An interval-balanced tournament design, IBTD$(9,2)$

0,1	4,5	0,8	3,4	7,8	1,4	5,8	2,4	3,8	0,4	1,8	0,5	2,8	3,5	4,8	1,5	6,8	2,5
2,3	6,7	1,2	5,6	0,2	3,6	0,7	1,6	5,7	2,6	3,7	4,6	1,7	0,6	2,7	0,3	4,7	1,3

the columns from c_1 to c_2 contains $2c(S_T+1)$ distinct elements, and thus

$$S_T \leq \frac{v}{2c} - 1.$$

If equality holds, then the set of elements in columns i and $i+S_T+1$ must be the same for all $1 \leq i \leq v(v-1)/2c - S_T - 1$. Thus, the sets of elements in each column 1 through S_T+1 partition V and no two points from different partitions can appear together anywhere else in the design, a contradiction to the definition of a tournament design. □

Definition 6.3 (interval-balanced tournament design). An *interval-balanced tournament design, IBTD(v,c), is a TD(v,c) with $S_T = s(v,c)$.*

An IBTD$(v,1)$ is simply a $(2d,d)$-ordering for a BIBD$(v,1,2)$ which Simmons and Davis have constructed [47] (see Theorem 4.1 on p. 78). Rodney uses precisely these designs to show that an IBTD always exists.

Theorem 6.2 ([41]). *For any integers v and c satisfying the necessary conditions, there exists an IBTD(v,c).*

Proof. Partition an IBTD$(v,1)$ into $1\times c$ subarrays, transpose each one and concatenate these horizontally, preserving their relative order in the IBTD$(v,1)$. The result is a $c \times v(v-1)/2c$ array.

An IBTD$(v,1)$ has separation $\lfloor (v-3)/2 \rfloor$. We want to determine the separation, S_T, of the new array we have constructed. Suppose the two closest appearances of an element w in the newly constructed array are in cells (r_1,c_1) and (r_2,c_1+S_t+1), then the construction gives

$$(c-r_1)+c\cdot S_T+(r_2-1) \geq \left\lfloor \frac{v-3}{2} \right\rfloor.$$

This reduces to

$$S_T \geq \left\lceil \frac{\lfloor \frac{v-3}{2} \rfloor - 2(c-1)}{c} \right\rceil. \tag{6.4}$$

Inequality (6.4) can be analysed for the cases (1) $v \equiv 0 \pmod{2c}$, (2) $v \equiv \ell \pmod{2c}$ with ℓ non-zero and even, and (3) $v \equiv \ell \pmod{2c}$ with ℓ odd, to check that it meets the bound of Lemma 6.1. □

An example IBTD is given in Table 6.3.

Table 6.4 A court- and interval-balanced tournament design, CIBTD$(9,2)$

2,3	6,7	0,8	3,4	0,2	1,4	5,8	1,6	5,7	2,6	3,7	0,5	2,8	0,6	4,8	1,5	4,7	1,3
0,1	4,5	1,2	5,6	7,8	3,6	0,7	2,4	3,8	0,4	1,8	4,6	1,7	3,5	2,7	0,3	6,8	2,5

If a tournament design is both a CBTD(v,c) and an IBTD(v,c), it is called a **court and interval-balanced tournament design**, CIBTD(v,c). Rodney showed that the two pairs in the columns of an IBTD$(v,2)$ can always be suitably permuted to produce a CIBTD$(v,2)$.

Theorem 6.3 ([41]). *For all positive integers v and c satisfying the necessary conditions for a CBTD$(v,2)$, there exists a CIBTD$(v,2)$.*

Proof. The necessary conditions for existence of a IBTD(v,c) imply that when $c = 2$, $v \equiv 1 \pmod 4$. Let $v = 4\ell + 1$. Swap the pairs in column i for all $i \equiv \{1,2,\ldots,\ell,2\ell+1,3\ell+2,3\ell+3,\ldots,4\ell+1\} \pmod{2v}$. The court-balanced nature of the resulting tournament design is established by considering the appearance of four types of element: (1) $x \equiv 0 \pmod 2$, $x \neq 1, v-1$, (2) $x \in \{3, v-1\}$, (3) $x \equiv 1 \pmod 4$, $x \neq 1$, and (4) $x \equiv 3 \pmod 4$, $x \neq 3$. □

An example CIBTD is given in Table 6.4.

Rodney used the patterned starter defined on p. 106 to come close to the bound of Lemma 6.1 for $c > 2$. The resulting designs cannot be called CIBTDs because the separation is not maximum.

Theorem 6.4 ([41]). *Let $v \equiv 1 \pmod{2c}$. There exists a CBTD(v,c) with separation $S_T = s(v,c) - 1$.*

Rodney computationally found examples of CIBTD$(13,3)$, CIBTD$(16,3)$, CIBTD$(19,3)$ and CIBTD$(22,3)$. To summarize and rephrase the results in this section in configuration ordering language, IBTD(v,c)s and CIBTD(v,c)s are simply TD(v,c)s and CBTD(v,c)s, respectively, with separation $s = s(v,c)$. That is, every set of s consecutive partial 1-factors are pairwise edge disjoint, meaning their union is also a partial 1-factor. In general, TD(v,c)s and CBTD(v,c)s with some separation s are decompositions of K_v into c-factors with a linear ordering on the c-factors such that the union of any consecutive $(s+1)$-set of them is a $c(s+1)$-factor.

6.2.3 *Time Intervals Before Repeating an Opponent*

In the early 2000s, a tennis-playing colleague of Stevens posed a question regarding a tournament schedule for her tennis club:

> There are seven people in the group. Only six people play on a given day, two people on each of three courts. After 45 minutes, they change opponents. The ideal rotation will have each person playing the other six people evenly, an even distribution of weeks off for each player, and the longest time possible between playing the same person again.

When v is odd, Theorem 4.42 (see p. 109) gives a schedule which optimally solves this problem. Each of the $(v-1)$-cycles can be decomposed into two near 1-factors, both missing the same point, which represents the player with a bye round. The two near 1-factors represent the pairings of players for each of the two games in that round. The edge-disjointness condition guarantees that any pair of players will not play each other for at least $(v-3)/2$ weeks which is the best possible. In fact, each player will have exactly two opponents who they play after $(v-3)/2$ weeks and will avoid playing all other players for $(v-1)/2$ weeks. Therefore, there are a total of v pairs appearing at separation $(v-3)/2$; this is the minimum possible for a cyclically developed solution.

6.2.4 *Home and Away Balance and Carry-Over Balance*

Of the various notions of ordering the columns of a tournament design, the interval-balanced designs are the easiest to define and discuss. They also fit nicely into the general model of configuration ordering. The next two orderings—which attempt to balance home and away effects and carry-over effects—do not fit so nicely into configuration ordering language.

6.2.4.1 Home and Away Balance

In sports scheduling it is assumed that a competitor playing on their home court has an advantage over their competitors. In tournament designs, the home and away teams in each match can be represented by an ordering of the pairs in each column, with the first coordinate indicating the home team and the second coordinate indicating the away team. When v is odd, the home advantage can be balanced by requiring that every team play half of its games at home and half away. When v is even, home advantage is balanced by requiring that the number of home and away games for any team differ by one. Additionally, we can ask that the number of consecutive home or consecutive away games (disregarding bye rounds) be minimized. Achieving home and away balance is a question of ordering the blocks (rounds) of a tournament design to induce a (global) set system that has balance properties. When c is as large as possible, home and away balance can always be achieved.

Theorem 6.5 ([52]). *There exists a* $TD(2k-1,k-1)$ *with a home-away designation of each pair and a circular ordering of the near* 1*-factors such that no team plays consecutive home or away games (ignoring bye rounds).*

Proof. Put the pair $\{i,j\}$ for $i<j$ in round $i+j \pmod{2k-1}$. Order the pair (i,j) if $i+j$ is odd and (j,i) if $i+j$ is even. □

An example of a home and away balanced TD$(7,3)$ is given in Table 6.5.

Table 6.5 An ordering of a TD$(7,3)$ such that no team plays consecutive home or away games (indicated by the first and second coordinate of each pair, respectively). For further clarity, each game has been placed in the row of its home team

Home team \ Round	0	1	2	3	4	5	6
0	Bye	(0,1)		(0,3)		(0,5)	
1	(1,6)		Bye	(1,2)		(1,4)	
2	(2,5)		(2,0)		Bye	(2,3)	
3	(3,4)		(3,6)		(3,1)		Bye
4		Bye	(4,5)		(4,0)		(4,2)
5		(5,3)		Bye	(5,6)		(5,1)
6		(6,2)		(6,4)		Bye	(6,0)

Table 6.6 An ordering of the blocks (rounds) of a TD$(8,4)$ such that the fewest possible teams play consecutive home or away games (indicated by the first and second coordinate of each pair, respectively). For further clarity, each game has been placed in the row of its home team. A match with team ∞ is equivalent to a bye round in the corresponding TD$(7,3)$

Home team \ Round	0	1	2	3	4	5	6
0	$(0,\infty)$	(0,1)		(0,3)		(0,5)	
1	(1,6)		$(1,\infty)$	(1,2)		(1,4)	
2	(2,5)		(2,0)		$(2,\infty)$	(2,3)	
3	(3,4)		(3,6)		(3,1)		$(3,\infty)$
4			(4,5)		(4,0)		(4,2)
5		(5,3)			(5,6)		(5,1)
6		(6,2)		(6,4)			(6,0)
∞		$(\infty,4)$		$(\infty,5)$		$(\infty,3)$	

In a TD$(2k,k)$ there must be a total, over all teams, of at least $2k-2$ consecutive home or away games if the ordering of blocks (rounds) is linear and at least $2k$ consecutive home or away games if the blocks (rounds) are circularly ordered. A tournament achieving this bound can be constructed from the TD$(2k-1,k-1)$ of Theorem 6.5 by assigning team ∞ to play with each team in its bye round and alternating ∞'s home and away status by the parity of the round. See Table 6.6 for an example.

While de Werra proved the existence of these designs [52], both examples presented use an earlier construction of Freund [24] which has the required home and away balance property, although he did not recognize this fact.

6.2.4.2 Carry-Over Balance

Our final example of block orderings from tournament designs is not a configuration ordering but rather an induced set system ordering in the sense discussed in

Table 6.7 A tournament design TD(8,4) which is balanced with respect to the carry-over effect

0,3	0,4	0,5	0,6	0,7	0,1	0,2
1,4	1,3	1,7	1,2	1,5	2,6	1,6
2,7	2,5	2,4	3,5	2,3	3,4	3,7
5,6	6,7	3,6	4,7	4,6	5,7	4,5

Sect. 3.3.1. The **carry-over** effect attempts to recognize that, in a match between teams x and y, the previous opponent faced by x has an effect on team y and vice versa. If team x played team z in its match previous to playing team y, then team y is said to receive a carry-over effect from team z. Russell describes one possible explanation of the nature of this effect:

> Each team is considered to have an effect on its opponents which carries over to the next match. If team A meets team B in one match and team C in the next, then it is reasonable that team A's performance against team C will have been affected by team B. Particularly in body-contact sports, if team B is a strong, hard-playing side, then team A is likely to enter the match against team C bruised in both body and morale. Conversely, if team B is relatively weak, then team C can anticipate that team A will be confident and fit for their match. Team C is said to receive a 'carry-over effect' due to team B. [43]

A tournament is **balanced with respect to the carry-over effect** if the carry-over effect from each team is spread as evenly as possible [43]. In particular, no team should receive the carry-over effect from the same team more than once.

Theorem 6.6 ([28, 43]). *When $v = 2^m$ and $c = v/2$, there exists a TD(v,c) and a circular ordering of its columns such that every team receives the carry-over effect from every other exactly once.*

Proof. Let the set of teams be $\mathbb{F}_v = \{0, \alpha, \alpha^2, \ldots, \alpha^{v-1}\}$, where α is a primitive element of the field. Check that $S_i = \sum_{j=1}^{i} \alpha^i$ is distinct for all $0 \le i < v-1$ (with the convention that an empty sum is 0). Let a be the value that does not appear as an S_i. The tournament design is constructed with team x playing team $(x + a + S_k)$ in the kth round.

Now suppose that team x receives a carry-over effect from team y in round k. This means there is a team, z, such that

$$y = z + a + S_{k-1},$$

$$x = z + a + S_k.$$

Thus, $x - y = S_k - S_{k-1} = \alpha^k$, and this can only happen in exactly one round. □

An example of this construction is given in Table 6.7. Anderson has been able to construct examples of TDs with balanced carry-over effect for $v = 20$ and 22 [1]. Russell has been able to construct designs that are not perfectly balanced but are close, for $v = q^e + 1$, where q is a prime [43].

By permuting the columns of addition tables of cyclic groups, Anderson and Bailey were able to construct carry-over balanced tournaments for two divisions, each having an even number of teams, where no two teams from the same division

Table 6.8 A bipartite tournament design for two divisions of six teams each, balanced with respect to the carry-over effect

a,0	a,1	a,5	a,2	a,4	a,3
b,1	b,2	b,0	b,3	b,5	b,4
c,5	c,0	c,4	c,1	c,3	c,2
d,2	d,3	d,1	d,4	d,0	d,5
e,4	e,5	e,3	e,0	e,2	e,1
f,3	f,4	f,2	f,5	f,1	f,0

play each other; this is the bipartite case. These tournaments are related to row-complete Latin squares which will be discussed in Sect. 6.5.2. An example of a bipartite TD that is balanced with respect to carry-over effect is given in Table 6.8. Using terraces, Bedford et al. were able to construct an example for two divisions of 21 teams which is given in Table 6.9 [2].

In general, controlling the carry-over effect in sports scheduling is very similar to controlling the residual effect in statistical design of experiments. We will discuss the residual effect in Sect. 6.5.2.

6.2.5 *Other Ordering Requirements in Scheduling*

The chapter on tournament scheduling in the *CRC Handbook of Combinatorial Designs* contains additional discussion of scenarios in which some form of ordering is important [19]. These include National Football League (NFL) schedules where teams play each other twice in a season; the first game must occur in the first 8 weeks of the season, and the second game must occur in the last 8 weeks. This is a block (round) ordering requirement similar to that of the maximally separated twofold near Hamilton decompositions described in Sect. 4.3.3. The NFL schedule also has home and away constraints. Other practical examples include the Czech National Hockey League and the Czech National Soccer League which utilize home and away balance and carry-over balance.

6.3 Reliability Testing

Reliability testing of software and hardware systems suffers from combinatorial explosion as a function of the number of parameters in the system under testing or the number of values each parameter can take on. For example, to exhaustively test a system with 10 binary parameters requires $2^{10} = 1,024$ tests, but double either the number of parameters or the size of each parameter and over a million tests will be required. One of the common solutions to avoid this explosion is interaction testing where test suites are designed so that all t-wise combinations of interactions are guaranteed to be covered. The design theoretic objects utilized in this testing are covering arrays.

Table 6.9 A bipartite tournament design for two divisions of 21 teams each, balanced with respect to the carry-over effect

a,0	a,1	a,6	a,7	a,11	a,5	a,17	a,10	a,9	a,16	a,18	a,14	a,8	a,3	a,12	a,20	a,2	a,15	a,13	a,19	a,4
b,1	b,2	b,0	b,8	b,12	b,6	b,18	b,11	b,10	b,17	b,19	b,15	b,9	b,4	b,13	b,14	b,3	b,16	b,7	b,20	b,5
c,2	c,3	c,1	c,9	c,13	c,0	c,19	c,12	c,11	c,18	c,20	c,16	c,10	c,5	c,7	c,15	c,4	c,17	c,8	c,14	c,6
d,3	d,4	d,2	d,10	d,7	d,1	d,20	d,13	d,12	d,19	d,14	d,17	d,11	d,6	d,8	d,16	d,5	d,18	d,9	d,15	d,0
e,4	e,5	e,3	e,11	e,8	e,2	e,14	e,7	e,13	e,20	e,15	e,18	e,12	e,0	e,9	e,17	e,6	e,19	e,10	e,16	e,1
f,5	f,6	f,4	f,12	f,9	f,3	f,15	f,8	f,7	f,14	f,16	f,19	f,13	f,1	f,10	f,18	f,0	f,20	f,11	f,17	f,2
g,6	g,0	g,5	g,13	g,10	g,4	g,16	g,9	g,8	g,15	g,17	g,20	g,7	g,2	g,11	g,19	g,1	g,14	g,12	g,18	g,3
h,7	h,9	h,12	h,14	h,15	h,10	h,6	h,20	h,18	h,4	h,1	h,0	h,16	h,13	h,17	h,5	h,11	h,2	h,19	h,3	h,8
i,8	i,10	i,13	i,15	i,16	i,11	i,0	i,14	i,19	i,5	i,2	i,1	i,17	i,7	i,18	i,6	i,12	i,3	i,20	i,4	i,9
j,9	j,11	j,7	j,16	j,17	j,12	j,1	j,15	j,20	j,6	j,3	j,2	j,18	j,8	j,19	j,0	j,13	j,4	j,14	j,5	j,10
k,10	k,12	k,8	k,17	k,18	k,13	k,2	k,16	k,14	k,0	k,4	k,3	k,19	k,9	k,20	k,1	k,7	k,5	k,15	k,6	k,11
ℓ,11	ℓ,13	ℓ,9	ℓ,18	ℓ,19	ℓ,7	ℓ,3	ℓ,17	ℓ,15	ℓ,1	ℓ,5	ℓ,4	ℓ,20	ℓ,10	ℓ,14	ℓ,2	ℓ,8	ℓ,6	ℓ,16	ℓ,0	ℓ,12
m,12	m,7	m,10	m,19	m,20	m,8	m,4	m,18	m,16	m,2	m,6	m,5	m,14	m,11	m,15	m,3	m,9	m,0	m,17	m,1	m,13
n,13	n,8	n,11	n,20	n,14	n,9	n,5	n,19	n,17	n,3	n,0	n,6	n,15	n,12	n,16	n,4	n,10	n,1	n,18	n,2	n,7
o,14	o,18	o,17	o,0	o,2	o,20	o,12	o,5	o,1	o,8	o,9	o,7	o,4	o,19	o,6	o,10	o,15	o,11	o,3	o,13	o,16
p,15	p,19	p,18	p,1	p,3	p,14	p,13	p,6	p,2	p,9	p,10	p,8	p,5	p,20	p,0	p,11	p,16	p,12	p,4	p,7	p,17
q,16	q,20	q,19	q,2	q,4	q,15	q,7	q,0	q,3	q,10	q,11	q,9	q,6	q,14	q,1	q,12	q,17	q,13	q,5	q,8	q,18
r,17	r,14	r,20	r,3	r,5	r,16	r,8	r,1	r,4	r,11	r,12	r,10	r,0	r,15	r,2	r,13	r,18	r,7	r,6	r,9	r,19
s,18	s,15	s,14	s,4	s,6	s,17	s,9	s,2	s,5	s,12	s,13	s,11	s,1	s,16	s,3	s,7	s,19	s,8	s,0	s,10	s,20
t,19	t,16	t,15	t,5	t,0	t,18	t,10	t,3	t,6	t,13	t,7	t,12	t,2	t,17	t,4	t,8	t,20	t,9	t,7	t,11	t,14
u,20	u,17	u,16	u,6	u,1	u,19	u,11	u,4	u,0	u,7	u,8	u,13	u,3	u,18	u,5	u,9	u,14	u,10	u,2	u,12	u,15

Each row of a covering array corresponds to a single test to be run. The columns represent parameters of the system, and the entries represent values they may take. Covering arrays exist and have been studied for various relaxations of the definition that conform to realistic settings, including each column having its own value set, possibly with different sizes (mixed-level covering arrays), and different sizes of interactions being tested (mixed-strength covering arrays) [13, 14, 26].

Covering arrays are natural generalizations of orthogonal arrays and are useful in testing because orthogonal arrays do not exist for all parameters t, k and v, whereas covering arrays always exist, though one must determine their optimal size. Furthermore, uniform coverage of t-tuples which is required of orthogonal arrays for their utility in statistical analysis is not required in software testing, where the results—*pass* or *fail*—are not linearly varying measurements. Thus, only the "at least one" coverage of covering arrays is necessary. Finally, a covering array, $\mathrm{CA}(N;t,k,v)$, is guaranteed to only grow logarithmically as a function of k [14]. There have been numerous studies showing that coverage of tuples for reasonably small t gives good testing results. In the particular instance of software testing, for example, t-tuple coverage gives good code and path coverage [6, 17, 20, 30, 31, 56]. Covering arrays are useful in a wide variety of different applications including software and hardware testing [3, 17, 30–32, 45, 51], genomics [46] and material sciences [7].

Recent research has extended the covering array model yet again to include ordering the rows (test cases) according to various priorities from actual testing scenarios. The rows of a covering array correspond precisely to the blocks of the corresponding design, so this new research is definitively discussing ordering the blocks of a design. The notion of ordering here is less mathematically determined, and the research typically presents algorithms for finding orderings given real-world quantitative priorities or evidence that these orderings are beneficial in practice.

There are numerous reasons why an ordered (prioritized) test suite is advantageous. These include finding faults as early in the testing as possible, achieving a certain level of code/path/interaction coverage as early as possible, allowing recoding to begin as early as possible, achieving reliability confidence earlier, finding higher-risk faults earlier, finding faults related to new code earlier, reducing costs, and increasing the likelihood of greatest coverage if testing is prematurely halted [21, 22, 42].

Recall that an f-prioritized ordering is defined as follows. Let $(V,\mathscr{B})$ be a block design. Let $P\mathscr{B}$ be the set of all permutations of the block set and let f be a function from $P\mathscr{B} \to \mathbb{R}$. A f-prioritized ordering for $(V,\mathscr{B})$ is a $\pi \in P\mathscr{B}$ such that $f(\pi) \geq f(\pi')$ for all other $\pi' \in P\mathscr{B}$. The functions f that are used to define f-prioritized orderings for the blocks of a covering array are typically derived from test experience of a previous version, known errors, areas with greater code coverage requirements, and historical failure data [5, 42]. Recognizing the computational hardness of determining the optimal row to add over the space of all possible rows, Rothermel et al. employ the dynamically prioritized method given in Definition 3.11, where the function evaluates which of the remaining test cases uncovered the most not-yet-found faults in previous versions of the software.

Algorithm 6.1 An algorithm for constructing covering arrays with prioritized rows

```
start with an empty covering array
while uncovered pairs remain do
  for each parameter do
    compute all parameter interaction weights
    initialize a new row with all parameters empty
    while a parameter remains whose value is not fixed in this row do
      select a parameter, p, that has the largest parameter interaction weight
      compute interaction weights for each value of parameter p
      select the value, x, which offers the largest increase in weighted density
      fix parameter p to value x
    end while
  end for
  add row to covering array
end while
```

Acknowledging again that this is a heuristic and not guaranteed to be optimal, the authors give an example of where this method fails to produce an optimal ordering [42]. Their next ordering is a simple descending list of how many code lines each test case covers. They then put this same measure into Definition 3.11 and choose test cases greedily by the maximum number of not-yet-covered lines of code. Elbaum et al. extend these general methods to include coverage of functions, fault existence, combined probabilities of fault existence and exposure [21, 22]. Whittaker et al. discuss user- or specification-based profiles which can be used to prioritize test suites [53, 54].

In [4], Bryce and Colbourn give a general-purpose covering array generation algorithm which can accept input priorities as a weight function on parameters, parameter values or even specific pairs of values. The algorithm is an adaptation of their Deterministic Density Algorithm which was designed to construct covering arrays one row at a time without resorting to a full greedy search through all potential unused rows and still provide a logarithmic growth guarantee. In general, the algorithm proceeds by calculating the "density" of various values of parameters, where density is defined as a weighted average over all interactions that a value participates in. The procedure is given in Algorithm 6.1.

Any weight function the user wants can be used in this algorithm, and it will generate a new test suite with guaranteed pairwise coverage in an ordering biased towards the weighting. Qu et al. adapt this method to reorder an existing test suite rather than generating a new test suite from scratch [39, 40]. The only modification made to the algorithm is that the domain from which a new row is chosen is the set of those in the original array, rather than all possible rows.

In general, the empirical studies referenced in this section found that ordering test cases can have significant impact on the efficacy of testing. Rothermel et al.

found that even the computationally cheapest priority functions showed significant improvement in early fault detection rates [42]. Elbaum et al. observed that different priority functions can have varying effectiveness over different target programs [21], but subsequently gained insight into the factors affecting the variance [22]. Qu et al. note that regenerating test suites anew can give decreased fault detection compared to reordering an existing test suite [40]. Most studies found cost-benefit trade-offs of various kinds.

Finally, we note that Srikanth et al. have continued the work of Qu et al. One of the new orderings they investigate is based on the cost of reconfiguring the software under test. In their examples, changing the configuration could involve a costly (in terms of time) change such as installing a new operating system or more moderate costs such as reconfiguring a security or database suite on a fixed operating system. The relevant ordering in this example is a Gray code ordering for the rows of the array, preferentially changing database or security suite until a change in operating system is forced. It is interesting to note that the improvement in time efficiency of using the Gray code meant that these experiments finished the entire test suite and therefore found *all* the faults in less than half the testing time of any other method. As the authors point out, if the goal is to find faults as early as possible, then optimizing for the major cost in testing, namely, *time*, is superior to all other optimizations even if those methods order the covering array to maximize fault-finding in the first 30% of its rows. In other words, Srikanth et al. conclude that "we must consider [the] fault detection rate over time, not by the number of configurations tested" [49].

6.4 Group Testing

Group testing involves the search for "positive" items amongst a set of items by means of pooling items and testing the pools for the "positives". It is assumed that a pool will test positive if and only if it contains at least one positive item. If some upper bound is known on the number of possible "positives", then pooling can probably find the positives faster than testing each individual item. For example, if there are 32 items and it is known that exactly one item is positive, then the following procedure detects the positive item in five tests by testing pools. First, test some set of 16 items. If the test returns a positive, then the single positive item is amongst these 16 and the other 16 may be discarded; if the test returns a negative, then the positive item is amongst the remaining 16 and the tested set of 16 may be discarded. Now, from the selected set of 16, test a set of eight items, and keep or discard a set of eight according to the same procedure. From the eight, test four to obtain a set of four. From the four, test two to obtain a set of two items. Finally, test one of these two items. Either it is positive, or it is negative and the single remaining item is positive. This is an **adaptive** procedure where we can choose the composition of the next pool based on the results of tests on the previous pools.

In a **non-adaptive** procedure, the set of all pools to test must be specified in advance and the positive item(s) must be determined from the results of all the tests. In this scenario, five tests again suffice to find precisely one positive item amongst 32. First, label the items from 0 to 31. For $0 \leq i < 5$, the ith test pool consists of all the items which have a 1 in the ith position of the binary expansion of their label. The result of each test now determines the ith binary bit of the label of the positive item. If the ith test returns a positive, then the label of the positive item has a 1 in position i. This is the basis for a nice parlour trick where the performer hands six sheets, each of which has a list of numbers from $\mathbb{Z}_{64}$ on it, to an audience member and asks him to pick a secret number less than 64 and hand back only those sheets that contain his number. The performer, just glancing at the sheets, can reveal the secret number. The sheets simply consist of the pools given above for each of the six binary bits needed to express the numbers from 0 to 63 in binary. To determine the secret number, the performer simply adds up the first (lowest) number on each sheet to reconstruct the secret value [48].

There are many connections between group testing and combinatorial designs, but most do not involve ordering the blocks of the design [16]. There is one group testing scenario which does involve an ordering of the blocks of the relevant design. Let $C = \{c_1, c_2 \ldots, c_n\}$ be a set of items and suppose that for $i < j$, if c_i and c_j are positive, then c_ℓ is positive for all $i < \ell < j$. Suppose further that there are no more than d positive items. The set C is said to have the ***d*-consecutive positive property**. This is not as unrealistic as it sounds. For example, in the testing of short DNA clones for overlap with a set of segments from a given string of DNA, the consecutive positive property holds [12]. Adaptively, this testing for overlap can be done in $\log_2 n + \log_2 d + c$ pools, and thus tests, for a fixed constant c. However, it can be shown that non-adaptive methods require at least $O(d + \log_2 n)$ tests [12]. By forming subgroups of items, Colbourn was able to show that the problem for $d > 2$ can be solved with an algorithm for the $d = 2$ case plus an additional $2d$ tests to resolve the subgroups at the end. Colbourn then solved the 2-consecutive positive problem (i.e. finding the two positive items) with $\lceil \log_2 n \rceil + 3$ tests in the following way. First, given a list that has the 2-consecutive property, label the items by the first n words of a binary Gray code of order $\lceil \log_2 n \rceil$. For the pool p_i with $0 \leq i < \lceil \log_2 n \rceil$, we include any item which has a 1 in the ith position of the binary expansion of its Gray code label. There are three remaining pools, q_0, q_1 and q_2, to form. In pool q_j, include all items whose original label is $j \pmod 3$. The Gray code labelling has the property that for any consecutive pair of labels, the Boolean sum ($0 \vee 0 = 0$, $0 \vee 1 = 1 \vee 1 = 1$) of the labels is equal to precisely one of the two labels, and thus these Boolean sums are distinct except for possibly pairs $x_i \vee x_{i+1}$ and $x_{i+1} \vee x_{i+2}$. If at most two items are positive and they must be consecutive, then the results of the first $\lceil \log_2 n \rceil$ pools narrow the possibilities down to at most three, $\{x_i, x_{i+1}\}$, $\{x_{i+1}\}$ or $\{x_{i+1}, x_{i+2}\}$, and the last three pools can be used to distinguish between these. We see in this example the significance of the ordering: consecutive Boolean sums are (nearly) distinct and consecutive labels differ in a small way. Thus, this ordering has both a local, configuration-type property and a (global) induced set system property where the multiset of consecutive unions

Table 6.10 Three maximal 2-consecutive positive detectable matrices

$$\begin{bmatrix} 0&1\\ 1&0 \end{bmatrix} \quad \begin{bmatrix} 1&0&1&0&0&0&1&0\\ 1&0&0&1&0&1&0&0\\ 0&1&0&0&0&1&0&1\\ 0&1&1&0&1&0&0&0 \end{bmatrix} \quad \begin{bmatrix} 0&0&1&0&0&0&0&1&0&1&0&1&0&1&1\\ 0&1&0&0&0&1&1&0&1&0&0&1&1&0&0\\ 1&0&1&0&0&0&1&1&0&1&0&1&0&0&0&0\\ 0&0&0&1&0&1&0&1&0&0&0&1&1&0&0&1\\ 1&0&1&0&1&0&0&1&0&0&0&0&1&1&0 \end{bmatrix}$$

has repetitions tightly controlled. Before we go on, we will translate this concrete example into combinatorial design language. The pools are the point set of a design, and the items are the blocks. In this example, with a maximal number of items, the design would have v points and $b = 2^{v-3}$ blocks. A point x is on a block B if the item corresponding to B appears in the pool corresponding to x. In some papers on this subject, this is written as a $v \times b$ binary matrix with a 1 indicating the inclusion described above. The Boolean sum of two columns corresponds to the union of the corresponding blocks.

The optimum procedure using $m = \lceil \log_2 n \rceil + 1$ tests for solving the 2-consecutive positive problem was given by Müller and Jimbo [36]. For a system with b items distinguishable by testing v pools, they look for a sequence of b sets of a v-set with the property that the collection of all blocks together with the unions of all consecutive pairs of blocks is distinct. It is clear that such a sequence of sets will work because the outcomes of the pools will identify either the points on a single block (if there is only one positive item) or on the union of two consecutive blocks (if there are exactly two positive items), and the condition on the blocks and their consecutive unions means that we will be able to uniquely distinguish all such cases. Since the blocks and the consecutive unions are subsets of a v-set and the elements of this collection must be distinct, we have that $b \leq 2^{v-1}$. When $b = 2^{v-1}$, Müller and Jimbo call the sequence of pools a **maximal 2-consecutive positive detectable matrix**. They show that these always exist for $v \neq 3$ [36]. Examples are given in Table 6.10.

Theorem 6.7 ([36]). *If, for $v > 2$, there exists a $v \times 2^{v-1}$ maximal 2-consecutive positive detectable matrix, then there exists a $(v+2) \times 2^{v+1}$ maximal 2-consecutive positive detectable matrix.*

Proof. Let $H = [x_1, x_2, \ldots, x_b]$ be a $v \times 2^{v-1}$ maximal 2-consecutive positive detectable matrix. A $(v+2) \times 2^{v+1}$ maximal 2-consecutive positive detectable matrix is

$$G = \begin{bmatrix} 0 & 0 & x_n & x_{n-1} & x_{n-2} & \cdots & x_2 & x_1 & x_1 & x_2 & \cdots & x_{n-1} & x_n & x_n & x_{n-1} & \cdots & x_2 & x_1 & x_1 & x_2 & \cdots & x_{n-1} & x_n \\ 0 & 1 & 0 & 0 & 0 & \cdots & 0 & 0 & 1 & 1 & \cdots & 1 & 1 & 1 & 1 & \cdots & 1 & 1 & 0 & 0 & \cdots & 0 & 0 \\ 1 & 0 & 1 & 1 & 1 & \cdots & 1 & 1 & 0 & 0 & \cdots & 0 & 0 & 1 & 1 & \cdots & 1 & 1 & 0 & 0 & \cdots & 0 & 0. \end{bmatrix}$$

□

Corollary 6.1 ([36]). *There exists a $v \times 2^{v-1}$ maximal 2-consecutive positive detectable matrix for all positive $v \neq 3$.*

Table 6.11 A 4×6 maximal 2-consecutive positive detectable matrix with column weight two

$$\begin{bmatrix} 1&1&0&1&0&0\\ 0&1&0&0&1&1\\ 1&0&1&0&0&1\\ 0&0&1&1&1&0 \end{bmatrix}$$

Table 6.12 A 6×20 cyclic maximal 2-consecutive positive detectable matrix with column weight three

$$\begin{bmatrix} 1&1&1&0&0&0&1&1&0&0&1&1&0&1&1&1&0&0&0&0\\ 1&0&0&0&1&1&0&0&1&1&0&1&0&0&1&1&1&0&0&1\\ 0&0&1&1&0&0&0&1&0&1&1&0&1&0&0&1&1&1&0&1\\ 1&1&0&0&0&1&1&1&1&0&0&0&1&0&0&0&1&1&1&0\\ 0&0&0&1&1&1&1&0&0&0&1&1&0&1&0&0&0&1&1&1\\ 0&1&1&1&1&0&0&0&1&1&0&0&1&1&1&0&0&0&1&0 \end{bmatrix}$$

There are two aspects of the preceding solutions which are not ideal in practice. First, these codes cannot cope with any false-positive or false-negative tests [12]. Second, it is desirable that the number of pools any item is in be small compared to v and, preferably, be a fixed size [36]. Sagols et al. address the first problem in a manner similar to Colbourn, using a Gray code. They ask for an ordering of the fixed parity binary words of a given length such that for any two consecutive words, one dominates the other (i.e. is a super-set).

Theorem 6.8 ([44]). *Let $v \geq 4$ be an even number and $p \equiv v/2+1 \pmod 2$. There exists a cyclic sequence of all length-v parity p words such that for any pair of consecutive words, they are Hamming distance two and one dominates the other.*

Sagols et al. also show that for each v, the other parity class cannot have such a sequence. However, if only domination is required, then such sequences exist for all $v \geq 10$ for both parity classes.

Theorem 6.9 ([44]). *For $v \geq 10$ and parity class, p, there exists a cyclic sequence of all length-v parity p words such that for any pair of consecutive words, one dominates the other.*

Both of these codes give solutions for the 2-consecutive positive problem that can detect up to one false-positive or false-negative test.

Müller and Jimbo address the second property required in practice: that the blocks have constant weight. They call a $v \times \binom{v}{k}$ 2-consecutive positive detectable matrix with the property that all columns are of weight k a **maximal 2-consecutive positive detectable matrix with column weight k** (M2CPDM(v,k)). When the cyclic sequence of columns still has the 2-consecutive positive detectable property, it is called a **cyclic maximal 2-consecutive positive detectable matrix with column weight k** (CM2CPDM(v,k)). Examples are given in Tables 6.11 and 6.12.

Müller and Jimbo present several recursive constructions which combine to give the following theorem.

Theorem 6.10 ([36]). *For any positive v and any $1 \leq k \leq \lfloor v/2 \rfloor$, there exists a M2CPDM$(v,k)$ and, except for $(v,k) \in \{(2,1),(4,2)\}$, there exists a CM2CPDM$(v,k)$.*

Müller and Jimbo were the first to construct solutions that had both an error-detecting capability and fixed block size. In [37], Müller and Jimbo change terminology and talk in terms of cyclic sequences of the k-subsets of a v-set. A sequence $B_1, B_2, \ldots, B_{\binom{v}{k}}$ of k-subsets of a v-set is said to be a **cyclic sequence with distinct consecutive unions** (CSDU(v,k)) if the set of unions, $\{U_i = B_i \cup B_{i+1}\}$, are pairwise distinct. If all the unions have a fixed size ℓ, then we refer to a CSDU$(v,k|\ell)$, and if all have parity $p \pmod 2$, then we refer to a CSDU$(v,k|\mathscr{P}_p(v))$. Müller and Jimbo construct many such designs.

Theorem 6.11 ([37]). *There exists:*

- *A CSDU$(v,2|4)$ if and only if $v \geq 6$*
- *A CSDU$(v,3|5)$ if and only if $v \geq 8$*
- *A CSDU$(v,4|\mathscr{P}_0(v))$ if $v \geq 11$*
- *A CSDU$(v,5|\mathscr{P}_1(v))$ if $v \geq 12$*
- *A CSDU$(v,6|\mathscr{P}_0(v))$ if $v \geq 17$*
- *A CSDU$(v,7|\mathscr{P}_1(v))$ if $v \geq 19$.*

Since, in these constructions, the unions are always the same parity as k, the blocks and consecutive unions form a code with minimum distance two. In this context, a code is nothing more than a collection of blocks or, equivalently, their incidence vectors. The relevant distance metric is the cardinality of the symmetric difference of a pair of blocks as sets. As a result, the code is one-error detecting in the group testing application and also has the desirable property that every item is in a fixed number of pools (k). Müller and Jimbo point out that a CSDU$(2k+1,k|k+1)$ is a solution to the middle levels problem and is thus considered very hard. They then pose a number of research questions which we include in Sect. 6.5.4.

In order to correct errors, or detect more errors, a code with larger minimum distance is required. Momihara and Jimbo show that a sequence of the triples of a maximum packing design and their consecutive unions yield a minimum distance three code if consecutive triples are disjoint, and this disjointness implies that the consecutive unions are distinct. Thus, for $v \equiv 1,3 \pmod 6$, they offer an alternative proof of the existence of A_1-cyclic orderings for Steiner triple systems, but they also establish new results for the other modular classes of v.

Theorem 6.12 ([34]). *For all $v \geq 10$, there exists an optimal $2-(v,3,1)$ packing having an A_1-cyclic ordering.*

Finally, Momihara and Jimbo also extend their results to codes with minimum distance four which can correct one error and detect up to two errors. They start with a recursive result.

Lemma 6.2 ([33]). *1. If there exists a SQS(v) having an A_1'-cyclic ordering, then there exists a SQS$(2v)$ having an A_1'-cyclic ordering.*

2. If there exists a SQS(v) having an A_1'-cyclic ordering, then there exists a SQS$(3v-2)$ having an A_1'-cyclic ordering.

They exhibit SQSs of orders $v = 14$ and 16 which admit A'_1-cyclic orderings, allowing them to conclude the following.

Theorem 6.13 ([33]). *There exists an infinite family of SQSs which admit A'_1-cyclic orderings. Small v in this family include*

$$v = 14, 16, 28, 32, 40, 46, 56, 64, 80, 82, 92, 94, 112, 118, 128, 136, 160, 164, 166, 184, 188, 190, 224, 236, 238, 244, 256, 272, 274, 280, 320, 328, 332, 334, 352, 368, 376, 380, 382, 406, 448, 472, 476, 478, 488, 490, 496.$$

Again, we note that these sequences of blocks have two important properties. Consecutive blocks form some kind of local configuration: containment, disjointness, or Hamming distance two. And the set of consecutive unions also satisfies some kind of (global) induced set system property, such as distinctness. In the cases of packings of pairs by triples or Steiner quadruple systems, it turns out that the local property and properties of the design combine to imply the global property.

6.5 Statistical Experiments

As design theory has its roots in the design of statistical experiments, it is not surprising to find that ordering the blocks of designs has applications in this area.

6.5.1 *Paired Comparisons*

Paired comparison refers to any process of comparing the items from a set two at a time. It is used for the study of preferences in marketing, transport, economics, psychology, voting systems and social choice functions [25,55]. Performing several pairwise comparisons is believed to be a more realistic request than asking a subject to rank a large set. It is known that the order in which pairs are offered to a subject can have an impact on their expressed preferences. Additionally, when it is not possible to offer the two choices simultaneously, the order in which the items within a pair are presented is also relevant, for example, in the tasting of two drinks. In order to control these influences, it is desirable to have (1) every item compared with every other item, (2) consecutive presentations of the same item maximally separated, and (3) every item equally often presented first within a pair as presented second.

If there is a single experimental subject (e.g., a person) evaluating all the items, then Simmons and Davis's pair designs achieve precisely these three objectives [47] (see Theorem 4.1 on p. 78). Goos and Großmann discuss more complicated designs of this type in [25].

6.5.2 Serial Treatments

If treatments are being applied in a serial and periodic manner to one or more test subjects, then controlling for the affect of the previous treatment on the next is important. In the case of a single test subject, such designs have been studied for several decades [29, 38, 50].

Definition 6.4 (type 1 serially balanced sequence). Let v and λ be positive integers. A type 1 serially balanced sequence of order v and index λ, SBS1(v,λ), is a sequence of length λv^2+1 such that the following properties hold:

1. The first and last elements are the same.
2. The first element appears $\lambda v+1$ times, all other elements appear λv times.
3. Each of the v^2 ordered pairs of elements appears λ times among the λv^2 pairs of consecutive elements.
4. After deleting the first element and partitioning the remaining sequence into λv v-sets, each element appears precisely once in each of the sets.

Definition 6.5 (type 2 serially balanced sequence). Let v and λ be positive integers. A type 2 serially balanced sequence of order v and index λ, SBS2(v,λ), is a sequence of length $\lambda v(v-1)+1$ such that the following properties hold:

1. The first and last elements are the same.
2. The first element appears $\lambda(v-1)+1$ times; all other elements appear $\lambda(v-1)$ times.
3. Each of the $v(v-1)$ ordered pairs of distinct elements appears λ times amongst the $\lambda v(v-1)$ pairs of consecutive elements.
4. After deleting the first element and partitioning the remaining sequence into $\lambda(v-1)$ v-sets, each element appears precisely once in each of the sets.

In block design terminology, these are simply BIBD$(v,2,\lambda)$s where the points within a block are ordered *and* the blocks themselves are ordered so that consecutive blocks intersect in precisely one point. In particular, there are rank two (where this rank refers to block size rather than index) Ucycles for these designs. An example SBS1$(3,2)$ is $1,1,2,3,3,1,2,2,3,1,1,3,2,2,1,3,3,2,1$, which represents the following ordering of blocks (where ordered blocks are represented by the use of $(,)$ instead of $\{,\}$):

$$(1,1),(1,2),(2,3),(3,3),(3,1),(1,2),(2,2),(2,3),(3,1),$$
$$(1,1),(1,3),(3,2),(2,2),(2,1),(1,3),(3,3),(3,2),(2,1).$$

It is easy to construct SBSs from existing SBSs by simply repeating the original sequence and identifying the first and last elements in consecutive copies; these will be SBSs on the same elements but with higher index.

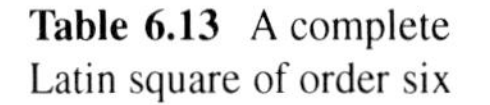

Table 6.13 A complete Latin square of order six

1	2	0	3	5	4
2	3	1	4	0	5
0	1	5	2	4	3
3	4	2	5	1	0
5	0	4	1	3	2
4	5	3	0	2	1

Lemma 6.3 ([50]). *If there exists an SBSi(v,λ), then there exists an SBSi$(v,m\lambda)$, for every integer $m \geq 1$, where $i = 1$ or 2.*

The existence of SBS2$(v,1)$s is completely determined.

Theorem 6.14 ([50]). *There exists an SBS2$(v,1)$ for all $v \geq 4$.*

Not much is known about SBS1$(v,1)$s, but general constructions exist for $\lambda = 2$.

Theorem 6.15 ([50]). *There exists an SBS1$(v,2)$ for all $v \geq 4$.*

In the case where there are several test subjects, statisticians use a specialized Latin square to determine treatment application.

Definition 6.6 (row complete). A Latin square is said to be row complete if every ordered pair of distinct symbols appears in adjacent positions precisely once amongst the rows of the square.

Definition 6.7 (column complete). A Latin square is said to be column complete if every ordered pair of distinct symbols appears in adjacent positions precisely once amongst the columns of the square.

When a Latin square is both row and column complete, it is called **complete**. A Latin square where every *unordered* pair appears twice in the rows and twice in the columns is called **quasi-complete**. Row-complete Latin squares are used in much the same way as SBSs except that each row represents the treatment schedule for a different subject. The columns represent some linearly varying factor, like time. A complete or quasi-complete Latin square is used when there are two linearly varying factors which need to be controlled [50].

Theorem 6.16 ([50]). *There exists a complete Latin square of order $2m$ for every integer $m \geq 1$.*

Proof. Let the first row and column of the square be

$$1 \;\; 2 \;\; 0 \;\; 3 \;\; (2m-1) \;\; 4 \;\; \cdots \;\; m \;\; (m+2) \;\; (m+1).$$

For the remaining entries of the square, set $\ell_{i,j} = \ell_{i,1} + \ell_{1,j} - 1 \pmod{2m}$. □

An example of a complete Latin square is given in Table 6.13.

For odd orders, the best that can currently be achieved is quasi-completeness.

Theorem 6.17 ([50]). *There exists a quasi-complete Latin square of order $2m+1$ for every integer $m \geq 1$.*

Table 6.14 A 4×16 array on four symbols with directional balance

1	4	2	3	2	1	3	4	2	1	3	4	1	4	2	3
4	3	1	2	1	4	2	3	1	4	2	3	4	3	1	2
2	1	3	4	3	2	4	1	3	2	4	1	2	1	3	4
3	2	4	1	4	3	1	2	4	3	1	2	3	2	4	1

Proof. The construction is the same as for complete Latin squares of even order. Let the first row and column of the square be

$$1\ 2\ 0\ 3\ 2m\ 4\ \cdots\ m\ (m+3)\ (m+1)\ (m+2).$$

For the remaining entries of the square, set $\ell_{i,j} = \ell_{i,1} + \ell_{1,j} - 1 \pmod{2m+1}$. □

The completeness concept can be extended to non-square arrays.

Definition 6.8 (directional balance). An $r \times c$ array, with entries from a v-set, in which each symbol appears equally often and never adjacent to itself in either rows or columns, is said to have directional balance if every ordered pair of distinct symbols appears adjacent equally often in rows and equally often in columns.

The balance is said to be **non-directional** if each *unordered* pair of distinct symbols appears adjacent equally often in rows and equally often in columns.

Table 6.14 shows an array with directional balance for $v = r = 4$ and $c = 16$.

Little is known in general regarding existence of balanced arrays. The most general result known involves non-directional balance.

Theorem 6.18 ([50]). *If $v = 4$, $r = 6s$ and $c = 6d$, with integers $s, d \geq 1$, then there exists an $r \times c$ array with non-directional balance.*

Nair has defined a variant for triples (instead of pairs) called a **serially balanced sequence of order v and index λ balanced for pairs of residuals**. These sequences must contain each of the distinct ordered triples exactly λ times and so will have length $\lambda v(v-1)(v-2)+2$. Nair proved that these sequences always exist for all v and λ [38]. These are rank three Ucycles for the λ distinct ordered triples of a v-set.

Throughout this section we have used the balance terminology to discuss the serial application of treatments with control of residual effects carried over from the previous treatment to the next. Such designs also appear in the literature under the name **change-over designs** [18].

6.5.3 Directional Seed Orchard Designs

In experiments involving collecting seed from a wind-pollinated species, all plants in a plot of land (block) are pollinators, but seed is only collected from plants not on the border of the plot [50]. If the block size is three, then this scenario is represented by ordering the points on each block so that every point appears equally often in the middle of a block and, when all blocks having the point x in the middle are considered, the set consisting of the first point in these blocks contains every treatment, except x, equally often. This is an example of an induced set system ordering (see Sect. 3.3.1). These designs are used primarily when wind direction is generally consistent [50].

Theorem 6.19 ([50]). *When $v = 6t+1$ is a prime, there exists a directional seed orchard design of order v and block size three.*

6.5.4 Neighbour Designs

Neighbour designs are cycle systems where the definition is relaxed to allow for a point to appear multiple times in a cycle, that is, neighbour designs could be called "tour systems". These designs are used in serology where different cultures are placed in a circle on a plate with growth media. Neighbours will interact with each other and possibly with some antiserum placed in the centre of the plate [50]. This is primarily a notion of (cyclically) ordering the points within a block, but Street and Street point out that if the cycles are put in sequence such that any two consecutive cycles have at least one point in common, then the cycles can be combined into longer cycles very much as was done in the proof of existence of rank two Ucycles for cyclic BIBDs (see Sect. 5.3.2). Constructions for some general neighbour designs are known.

Theorem 6.20 ([50]). *For the following pairs of (v,k), there exists a neighbour design of k-cycles from a v-set such that its blocks can be ordered in linear sequence with consecutive blocks not disjoint:*

- $(2k+1,k), k \geq 3$
- $(2^{i+1}\ell+1, 2\ell), \ell \geq 2$ *and* $i \geq 1$
- $(8m\ell+1, 4\ell), m, \ell \geq 1$

Exercises and Problems

Exercise 6.1. Prove that in a $\mathrm{TD}(2k,k)$ there must be at least $2k-2$ consecutive home games or $2k-2$ consecutive away games if the ordering of blocks is linear and at least $2k-1$ if the blocks are circularly ordered.

Exercise 6.2. Derive bounds on the number of consecutive (ignoring bye rounds) home or away games for a $\mathrm{TD}(v,c)$ where $v < (v-1)/2$.

Exercise 6.3. Can the $\mathrm{TD}(8,4)$ in Table 6.7 be made court-balanced? What about home and away balanced?

Exercise 6.4. Prove that there is no maximal 2-consecutive positive detectable matrix for $v = 3$.

Exercise 6.5. What is the largest b such that a $3 \times b$ 2-consecutive positive detectable matrix exists?

Exercise 6.6. Prove that a $v \times 2^{v-1}$ maximal 2-consecutive positive detectable matrix cannot have its columns *cyclically* sequenced, namely, that if we also consider the union of the first and last column (block), the 2-consecutive positive detectable property (see p. 191) fails.

Exercise 6.7. Show that for n even and $p \equiv n/2 \pmod 2$ there is no cyclic sequence of the length-n parity p binary words such that any pair of consecutive words are Hamming distance two and one dominates the other.

Exercise 6.8. For what values of $n < 10$ and which parity classes p, does there not exist a cyclic sequence of the length-n parity p binary words such that for any pair of consecutive words, one dominates the other?

Problem 6.1. If the ordering of the columns in the definition of an IBTD is considered to be cyclic rather than linear, what is the upper bound on S_T? Can the optimum be constructed?

Problem 6.2. Find more constructions of CIBTD(v,c)s for $c \geq 3$.

Problem 6.3. Generalize the definitions and bounds for IBTD(v,c)s to GIBTD (v,c,k)s. Find constructions meeting or coming close to these bounds.

Problem 6.4. Generalize the definitions and bounds for CIBTD(v,c)s to GCIBTD (v,c,k)s. Find constructions meeting or coming close to these bounds.

Problem 6.5. Construct TD(v,c)s with orderings meeting, or coming close to, the bounds derived in answering Exercise 6.2.

Problem 6.6. One of the ways of generalizing the home and away status of a match to $k > 2$ teams is called an r-tournament. An ***r*-tournament** is a contest in which competitors meet r at a time, every r set occurs exactly once and each r-set has an ordering such that every competitor appears in every position an equal number of times (see Chap. VI.51 of [15]). What are the necessary conditions for the existence of an r-tournament? Note that the existence of a Ucycle for the r-subsets of an v-set is sufficient to show the existence of an r-tournament. Generalize this notion to a tournament where every pair of players appears in precisely λ r-sets. What are the necessary conditions? Can you generate any examples? What is the appropriate way to generalize the constraint that consecutive home or away games appear the least number of times possible? Is there a more natural way to generalize this concept?

Problem 6.7. Construct a TD$(2k,k)$, k not a power of two and $k \neq 20, 22$, that is balanced with respect to the carry-over effect.

Problem 6.8. Construct a bipartite tournament for two divisions of v teams, for odd $v \neq 21$, that is balanced with respect to the carry-over effect .

Problem 6.9. What is the natural generalization of balanced with respect to carry-over effect for tournaments with match size $k > 2$?

Problem 6.10. Is it possible to show that there exists a CSDU$(2k+2, k|k+2)$ for all k? If so, recursive constructions could potentially yield CSDU$(v, k|k+2)$s for all $v \geq 2k+2$.

Problem 6.11. Weaker than the question posed as Problem 6.10, is it possible to show that there always exists a $CSDU(2k+2, k|\mathscr{P}_{k \pmod 2}(v))$?

Problem 6.12. Can a Steiner quadruple system of order $3v-2u$ having an A_1'-cyclic ordering be constructed from a Steiner quadruple system of order v having an A_1'-cyclic ordering such that the larger SQS contains a subsystem of order v with its blocks consecutive in the order?

Problem 6.13. Does there exist a Steiner quadruple system of order v with an A_1'-cyclic ordering for every $v \geq 12$?

Problem 6.14. Investigate the construction of SBS1$(v, 1)$s.

Problem 6.15. Investigate the existence of complete Latin squares of odd order.

References

1. Anderson, I.: Balancing carry-over effects in tournaments. In: Combinatorial Designs and their Applications (Milton Keynes, 1997), vol. 403, pp. 1–16. Chapman & Hall/CRC, Boca Raton, FL (1999)
2. Bedford, D., Ollis, M.A., Whitaker, R.M.: On bipartite tournaments balanced with respect to carry-over effects for both teams. Discrete Math. **231**(1–3), 81–87 (2001)
3. Brownlie, R., Prowse, J., Phadke, M.S.: Robust testing of AT&T PMX/StarMAIL using OATS. AT&T Technical Journal **71**, 41–47 (1992)
4. Bryce, R., Colbourn, C.J.: The density algorithm for pairwise interaction testing. Software Testing, Verification, and Reliability **17**, 159–182 (2007)
5. Bryce, R.C., Colbourn, C.J.: Prioritized interaction testing for pairwise coverage with seeding and constraints. Inform. Software Tech. **48**, 960–970 (2006)
6. Burr, K., Young, W.: Combinatorial test techniques: table-based automation, test generation, and code coverage. In: Proceedings of International Conference on Software Testing Analysis and Review, San Diego, CA, pp. 503–513 (1998)
7. Cawse, J.N.: Experimental design for combinatorial and high throughput materials development. GE Global Research **29**, 769–781 (2002)
8. Cohen, M.B., Colbourn, C.J.: Ordering disks for double erasure codes. In: Proceedings of Symposium Parallel Algorithms and Architectures (SPAA01), Crete Island, Greece, pp. 229–236 (2001)
9. Cohen, M.B., Colbourn, C.J.: Optimal and pessimal orderings of Steiner triple systems in disk arrays. Theoret. Comput. Sci. **297**, 103–117 (2003)
10. Cohen, M.B., Colbourn, C.J.: Ladder orderings of pairs and RAID performance. Disc. Appl. Math. **138**(1), 35–46 (2004)
11. Cohen, M.B., Colbourn, C.J., Fronček, D.: Cluttered orderings for the complete graph. In: Lecture Notes in Computer Science, vol. 2108, pp. 420–431 (2001)
12. Colbourn, C.J.: Group testing for consecutive positives. Annals of Combin. **3**, 37–41 (1999)

13. Colbourn, C.J.: Combinatorial aspects of covering arrays. Matematiche (Catania) **59**(1–2), 125–172 (2004)
14. Colbourn, C.J.: Covering Arrays, chap. VI.10, pp. 361–365. In: Colbourn and Dinitz [15] (2007)
15. Colbourn, C.J., Dinitz, J.H. (eds.): Handbook of Combinatorial Designs, second edn. Chapman & Hall/CRC, Boca Raton, FL (2007)
16. Colbourn, C.J., Hwang, F.K.: Superimposed Codes and Combinatorial Group Testing, chap. VI.56, pp. 629–633. In: Colbourn and Dinitz [15] (2007)
17. Dalal, S.R., Jain, A., Karunanithi, N., Leaton, J.M., Lott, C.M., Patton, G.C., Horowitz, B.M.: Model-based testing in practice. In: Proceedings of International Conference on Software Engineering (ICSE 99), Los Angeles, CA, pp. 285–294 (1999)
18. Davis, A.W., Hall, W.B.: Cyclic change-over designs. Biometrika **56**, 283–293 (1969)
19. Dinitz, J.H., Fronček, D., Lamken, E.R., Wallis, W.D.: Scheduling a Tournament, chap. VI.51, pp. 591–606. In: Colbourn and Dinitz [15] (2007)
20. Dunietz, S., Ehrlich, W.K., Szablak, B.D., Mallows, C.L., Iannino, A.: Applying design of experiments to software testing. In: Proceedings of International Conference on Software Engineering (ICSE 97), Boston, Massachusetts, pp. 205–215 (1997)
21. Elbaum, S., Malishevsky, A.G., Rothermel, G.: Test case prioritization: a family of empirical studies. IEEE Trans. Software Eng. **28**(2), 159–182 (2002)
22. Elbaum, S., Rothermel, G., Kanduri, S.: Selecting a cost-effective test case prioritization technique. Software Q. J. **12**(2), 185–210 (2004)
23. Folkman, J., Fulkerson, D.R.: Edge colorings in bipartite graphs. In: Combinatorial Mathematics and its Applications (Proc. Conf., Univ. North Carolina, Chapel Hill, N.C., 1967), pp. 561–577. Univ. North Carolina Press, Chapel Hill, N.C. (1969)
24. Freund, J.E.: Classroom notes: round robin mathematics. Amer. Math. Monthly **63**(2), 112–114 (1956)
25. Goos, P., Großmann, H.: Optimal design of paired comparison experiments in the presence of within-pair order effects. Draft Manuscript (2011)
26. Hartman, A.: Software and hardware testing using combinatorial covering suites. In: Graph Theory, Combinatorics and Algorithms Edited by M. Charles Golumbic and I. Ben-Arroyo Hartman, pp. 237–266. Springer, New York (2005)
27. Hellerstein, L., Gibson, G.A., Karp, R.M., Katz, R.H., Patterson, D.A.: Coding techniques for handling failures in large disk arrays. Algorithmica **12**, 182–208 (1994)
28. Hwang, F.K.: How to design round robin schedules. In: Combinatorics, Computing and Complexity Edited by D. Zhu Du and G. Ding Hu (Tianjing and Beijing, 1988), Math. Appl. (Chinese Ser.), vol. 1, pp. 142–160. Kluwer Academic Publishers, Dordrecht (1989)
29. Jimbo, M.: Serially balanced designs related to balanced incomplete block designs. J. Japan Statist. Soc. **15**(2), 161–175 (1985)
30. Kuhn, D.R., Reilly, M.: An investigation of the applicability of design of experiments to software testing. In: Proceedings of 27th Annual NASA Goddard/IEEE Software Engineering Workshop, Greenbelt, Maryland, pp. 91–95 (2002)
31. Kuhn, D.R., Wallace, D.R., Gallo, A.M.: Software fault interactions and implications for software testing. IEEE Trans. Software Eng. **30**, 418–421 (2004)
32. Mandl, R.: Orthogonal Latin squares: an application of experiment design to compiler testing. Comm. ACM **28**, 1054–1058 (1985)
33. Momihara, K., Jimbo, M.: Some constructions for block sequences of Steiner quadruple systems with error-correcting consecutive unions. J. Combin. Des. **16**(2), 152–163 (2008)
34. Momihara, K., Jimbo, M.: On a cyclic sequence of a packing by triples with error-correcting consecutive unions. Util. Math. **78**, 93–105 (2009)
35. Müller, M., Adachi, T., Jimbo, M.: Cluttered orderings for the complete bipartite graph. Discrete Appl. Math. **152**(1–3), 213–228 (2005)
36. Müller, M., Jimbo, M.: Consecutive positive detectable matrices and group testing for consecutive positives. Discrete Math. **279**(1–3), 369–381 (2004)

37. Müller, M., Jimbo, M.: Cyclic sequences of k-subsets with distinct consecutive unions. Discrete Math. **308**(2–3), 457–464 (2008)
38. Nair, C.R.: Sequences balanced for pairs of residual effects. J. Amer. Statist. Assoc. **62**, 205–225 (1967)
39. Qu, X., Cohen, M.B., Rothermel, G.: Configuration-aware regression testing: an empirical study of sampling and prioritization. In: International Symposium on Software Testing and Analysis (ISSTA), Seattle, WA, pp. 75–85 (2008)
40. Qu, X., Cohen, M.B., Woolf, K.M.: Combinatorial interaction regression testing: a study of test case generation and prioritization. In: IEEE International Conference on Software Maintenance (ICSM), Maison Internationale, Paris, France, pp. 255–264 (2007)
41. Rodney, P.: Balance in tournament designs. Ph.D. thesis, University of Toronto, Toronto, ON (1993)
42. Rothermel, G., Untch, R.H., Chu, C., Harrold, M.J.: Prioritizing test cases for regression testing. IEEE Trans. Software Eng. **27**(10), 929–948 (2001)
43. Russell, K.G.: Balancing carry-over effects in round robin tournaments. Biometrika **67**(1), 127–131 (1980)
44. Sagols, F., Riccio, L.P., Colbourn, C.J.: Dominated error-correcting codes with distance two. J. Combin. Des. **10**(5), 294–302 (2002)
45. Seroussi, G., Bshouty, N.H.: Vector sets for exhaustive testing of logic circuits. IEEE Trans. Inform. Theory **34**(3), 513–522 (1988)
46. Shasha, D.E., Kouranov, A.Y., Lejay, L.V., F., C.M., Coruzzi, G.M.: Using combinatorial design to study regulation by multiple input signals: a tool for parsimony in the post-genomics era. Plant Physiology **127**, 1590–1594 (2001)
47. Simmons, G.J., Davis, J.A.: Pair designs. Comm. Statist. **4**, 255–272 (1975)
48. base sixteen.org: binary magic trick. http://cse4k12.org/binary/magic_trick.html. Accessed 23 Sept 2011
49. Srikanth, H., Cohen, M.B., Qu, X.: Reducing field failures in system configurable software: cost-based prioritization. In: International Symposium on Software Reliability Engineering (ISSRE), Mysuru, India, pp. 61–70 (2009)
50. Street, A.P., Street, D.J.: Combinatorics of Experimental Design. Oxford Science Publications. The Clarendon Press Oxford University Press, New York (1987)
51. Tang, D.T., Chen, C.L.: Iterative exhaustive pattern generation for logic testing. IBM Journal Research and Development **28**, 212–219 (1984)
52. de Werra, D.: Some models of graphs for scheduling sports competitions. Discrete Appl. Math. **21**(1), 47–65 (1988)
53. Whittaker, J.A., Poore, J.H.: Markov analysis of software specifications. ACM Trans. Software Eng. Meth. **2**(1), 93–106 (1993)
54. Whittaker, J.A., Thomason, M.G.: A Markov chain model for statistical software testing. IEEE Trans. Software Eng. **20**(10), 812–824 (1994)
55. Wikipedia: Pairwise comparison — wikipedia, the free encyclopedia. http://en.wikipedia.org/w/index.php?title=Pairwise_comparison&oldid=408248092. Accessed 21 June 2011
56. Williams, A.W., Probert, R.L.: A measure for component interaction test coverage. In: Proceedings of ACS/IEEE International Conference on Computer Systems & Applications, Beirut, Lebanon, pp. 301–311 (2001)

Index

M. Dewar and B. Stevens, *Ordering Block Designs: Gray Codes, Universal Cycles and Configuration Orderings*, CMS Books in Mathematics,
DOI 10.1007/978-1-4614-4325-4, © Springer Science+Business Media New York 2012